How Charts Lie

ALSO BY ALBERTO CAIRO

The Functional Art

The Truthful Art

How Charts Lie

Getting Smarter about Visual Information

Alberto Cairo

W. W. NORTON & COMPANY
Independent Publishers Since 1923

Printed in the United States of America
First Edition

For information about special discounts for bulk purchases, please contact W. W. Norton Special Sales at specialsales@wwnorton.com or 800-233-4830

Manufacturing by LSC Communications Harrisonburg
Book design by Daniel Lagin
Production manager: Lauren Abbate

ISBN 978-1-324-00156-0

W. W. Norton & Company, Inc., 500 Fifth Avenue, New York, N.Y. 10110
www.wwnorton.com

W. W. Norton & Company Ltd., 15 Carlisle Street, London W1D 3BS

1 2 3 4 5 6 7 8 9 0

The world cannot be understood without numbers.
And it cannot be understood with numbers alone.

—HANS ROSLING, *FACTFULNESS* (2018)

Freedom depends upon citizens who are able to make a distinction between what is true and what they want to hear. Authoritarianism arrives not because people say that they want it, but because they lose the ability to distinguish between facts and desires.

—TIMOTHY SNYDER, *THE ROAD TO UNFREEDOM* (2018)

To my parents

Contents

Prologue

A World Brimming with Charts

This is a book about how the many charts—tables, graphs, maps, diagrams—we see every day through TV, newspapers, social media, textbooks, or advertising deceive us.

We've all heard the old idiom "A picture is worth a thousand words." My hope is that you'll soon stop using it unless you append it with this: "*if you know how to read it*." Even common charts such as maps and bar graphs can be ambiguous or, worse, incomprehensible.

This is worrying because numbers are very persuasive, and so are charts, because we associate them with science and reason. Numbers and charts look and feel objective, precise, and, as a consequence, seductive and convincing.[1]

Politicians, marketers, and advertisers throw numbers and charts at us with no expectation of our delving into them: the average family will save $100 a month thanks to this tax cut; the unemployment rate is at 4.5%, a historic low, thanks to our stimulus package; 59% of Americans disapprove of the president's performance; 9 out of 10 dentists recommend our toothpaste; there's a 20% chance of rain today; eating more chocolate may help you win the Nobel Prize.[2]

The moment we turn on the TV, open a newspaper, or visit our favorite social media network, we are bombarded with flashy charts. If you have a job, your performance has likely been measured and shown through graphics. You may have designed them yourself to insert into slides for a class or business presentation. Some authors with a taste for the hyperbolic even talk about a "tyranny of numbers" and a "tyranny of metrics," referring to the pervasiveness of measurement.[3] As people in contemporary societies, we are easily seduced by our numbers and by the charts that represent them.

Charts—even those not designed with ill intent—can mislead us. However, they can also tell us the truth. Well-designed charts are empowering. They enable conversations. They imbue us with X-ray vision, allowing us to peek through the complexity of large amounts of data. Charts are often the best way to reveal patterns and trends hidden behind the numbers we encounter in our lives.

Good charts make us smarter.

But before that happens, we need to get used to perusing them with attention and care. Instead of just *looking* at charts, as if they were mere illustrations, we must learn to *read* them and *interpret* them correctly.

Here's how to become a better chart reader.

How Charts Lie

Introduction

On April 27, 2017, President Donald J. Trump sat with Reuters journalists Stephen J. Adler, Jeff Mason, and Steve Holland to discuss his accomplishments in his first 100 days in office. While talking about China and its president, Xi Jinping, Trump paused and handed the three visitors copies of a 2016 electoral map:[1]

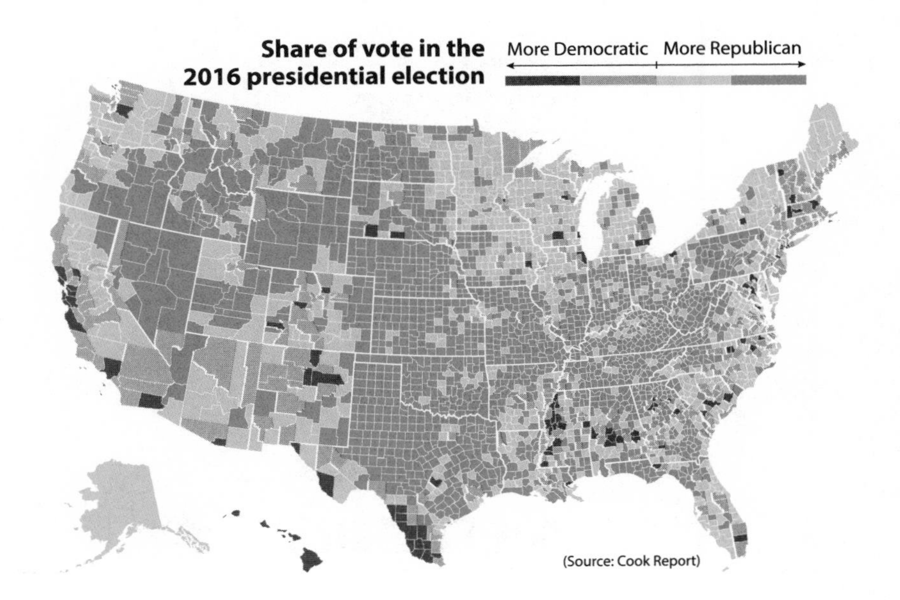

The president then said, "Here, you can take that, that's the final map of the numbers. It's pretty good, right? The red is obviously us."

When I read the interview, I thought that it was understandable President Trump was so fond of that map. He won the 2016 election despite most forecasts, which gave him between 1% and 33% chances of succeeding; a Republican establishment that distrusted him; a bare-bones campaign that was often in disarray; and numerous controversial remarks about women, minorities, the U.S. intelligence services, and even veterans. Many pundits and politicians predicted Trump's demise. They were proved wrong. He seized the presidency against all odds.

However, being victorious isn't an excuse to promote misleading visuals. When presented alone and devoid of context, this map can be misleading.

The map appeared in many other places during 2017. According to The Hill,[2] White House staffers had a large, framed copy of it hanging in the West Wing. The map was also regularly touted by conservative media organizations, such as Fox News, Breitbart, and InfoWars, among others. Right-wing social media personality Jack Posobiec put it on the cover of his book, *Citizens for Trump*, which looks similar to this:

I've spent the last two decades making charts and teaching others how to design them. I'm convinced that anyone—including you, reader—can learn how to read and even *create* good graphics, so I'm usually happy to offer my free and constructive advice to whoever wants to take it. When I saw Posobiec's book on social media, I suggested that he needed to change either the title or the map, as the map doesn't show what the book title says.

The map is misleading because it's being used to represent the *citizens* who voted for each candidate, but it doesn't. Rather, it represents *territory*. I suggested that Posobiec either change the graphic on the cover of his book to better support the title and subtitle, or change the title to *Counties for Trump*, as that is what the map truly shows. He ignored my advice.

Try to estimate the proportion of each color, red (Republican) and grey (Democratic). Roughly, 80% of the map's surface is red and 20% is grey. The map suggests a triumph by a landslide, but Trump's victory wasn't a landslide at all. The popular vote—Posobiec's "citizens"—was split nearly in half:

Share of the popular vote in the 2016 presidential election

Donald Trump **46.1%** 62,984,825 votes
Hillary Clinton **48.2%** 65,853,516 votes
Other candidates **5.7%**

We could be even pickier and point out that turnout in the election was around 60%;[3] more than 40% of eligible voters didn't show up at the polls. If we do a chart of *all eligible voters*, we'll see that the citizens who voted for each of the major candidates were a bit less than a third of the total:

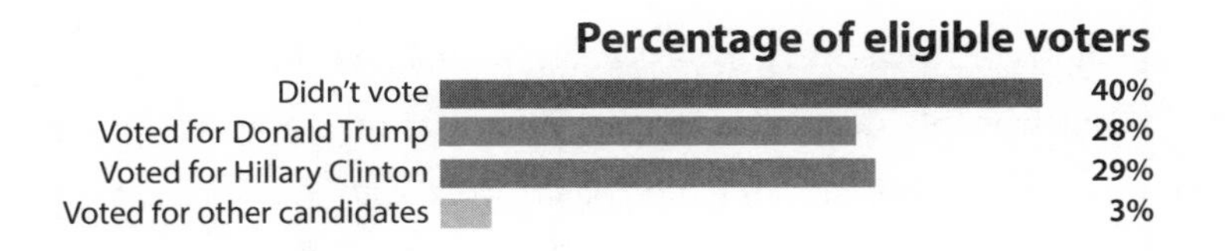

And what if we count *all* citizens? There are 325 million people in the United States. Of those, around 300 million are citizens, according to the

Kaiser Foundation. It turns out that "Citizens for Trump" or "Citizens for Clinton" are just a bit more than one-fifth of all citizens.

Critics of President Trump were quick to excoriate him for handing out the county-level map to visitors. Why count the square miles and ignore the fact that many counties that went for Trump (2,626)[4] are large in size but sparsely populated, while many of the counties where Clinton won (487) are small, urban, and densely populated?

That reality is revealed in the following map of the continental U.S., designed by cartographer Kenneth Field. Each dot here represents a voter—grey is Democratic and red is Republican—and is positioned approximately—but not exactly—where that person voted. Vast swaths of the U.S. are empty:

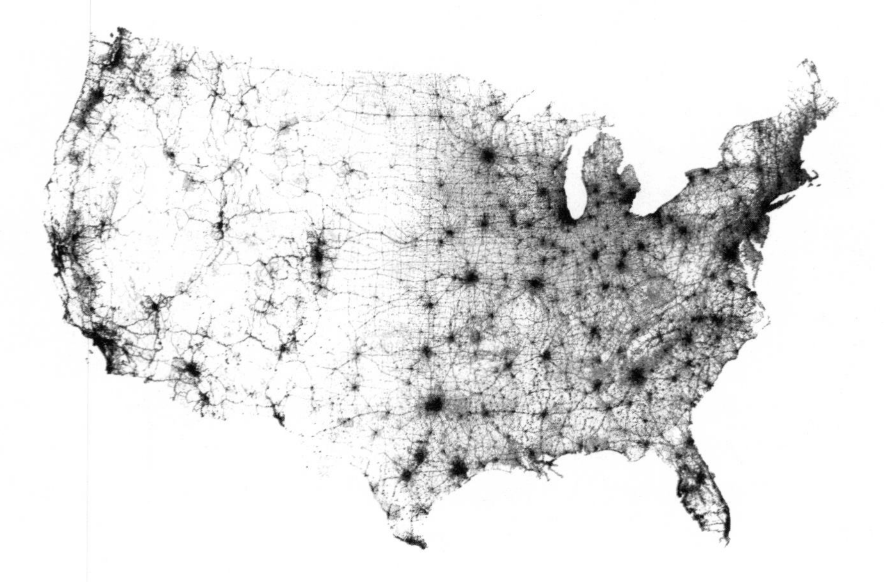

As someone who strives to keep a balanced media diet, I follow people and publications from all ideological stripes. What I've seen in recent years makes me worry that the increasing ideological polarization in the U.S. is also leading to a divide on chart preferences. Some conservatives I read love the county-level map President Trump handed out to reporters. They constantly post it on their websites and social media accounts.

Liberals and progressives, on the other hand, prefer a bubble map proposed by *Time* magazine and other publications.[5] In it, bubbles are sized in proportion to the votes received by the winning candidate in each county:

Both conservatives and liberals laugh at the other side's stupidity. "How can you tweet *that* map? Don't you see that it distorts the results of the election?"

This is no laughing matter. Both sides in this debate are throwing different charts at each other because we all often use information to reinforce our beliefs: conservatives love to convince themselves of a crushing victory in the 2016 election; liberals console themselves by emphasizing Hillary Clinton's larger share of the popular vote.

Liberals are correct when they claim that the colored county map isn't an adequate representation of the *number* of votes each candidate received, but the bubble map favored by liberals is also faulty. By showing only the votes for the winning candidate in each county, this chart ignores those received by the *losing* candidate. Plenty of people voted for Hillary Clinton in conservative regions. Many voted for Donald Trump in very progressive ones.

Kenneth Field's map or the pair of maps below may be a better choice if what we care about is the popular vote. There are many more visible red bubbles (votes for Trump) than grey bubbles (votes for Clinton), but the fewer grey ones are often much bigger. When these maps are put side by side, it's easier to see why the election was decided by a relatively small number of votes in a handful of states; if you add up the area of all red bubbles and the area of all grey bubbles, they are roughly the same:

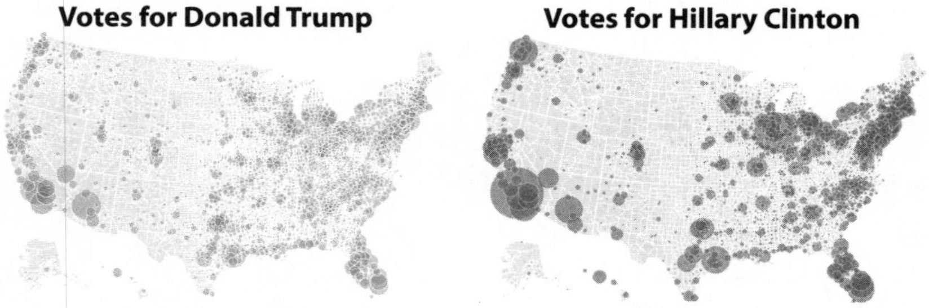

Votes for Donald Trump **Votes for Hillary Clinton**

Bubble size is proportional to the number of votes per county

Having said this, both conservatives and liberals are missing the point. What makes you win a presidential election in the United States is neither the territory you control, nor the number of people you persuade to vote for you *nationally*. It's the Electoral College and its 538 electors. To win, you need the support of at least 270 of electors.

Each state has a number of these folks equal to its congressional representation: two senators plus a number of representatives in the House that varies according to the state's population. If you are a small state with the fixed number of senators (two per state) plus one representative in the House, you are allotted three electors.

Small states often have more electors based on their populations than what pure arithmetic would give them: the minimum is three electors per state, no matter how small the population of that state is.

Here's how you receive the support of a state's electors: with the exception of Nebraska and Maine, the candidate who wins even a razor-thin advantage in a state's popular vote over his or her opponents is supposed to receive the support of *all* that state's electors.

In other words, once you've secured at least one more vote than any of your opponents, the rest of the votes you receive in that state are useless. You don't even need a majority, just a *plurality*: if you get 45% of the popular vote in one state, but your two opponents get 40% and 15%, you'll receive all the electoral votes from that state.

Trump got the support of 304 electors. Clinton, despite winning the national popular vote by a margin of three million and getting tons of support in populous states like California, received only 227. Seven electors went rogue, voting for people who weren't even candidates.

Therefore, if I ever got elected president—which is an impossibility, since I wasn't born in the U.S.—and I wanted to celebrate my victory by printing out some charts, framing them, and hanging them on the walls of my White House, it would be with the ones below. They are focused on the figures that really matter—neither the number of counties, nor the popular vote, but the number of *electoral* votes received by each candidate:

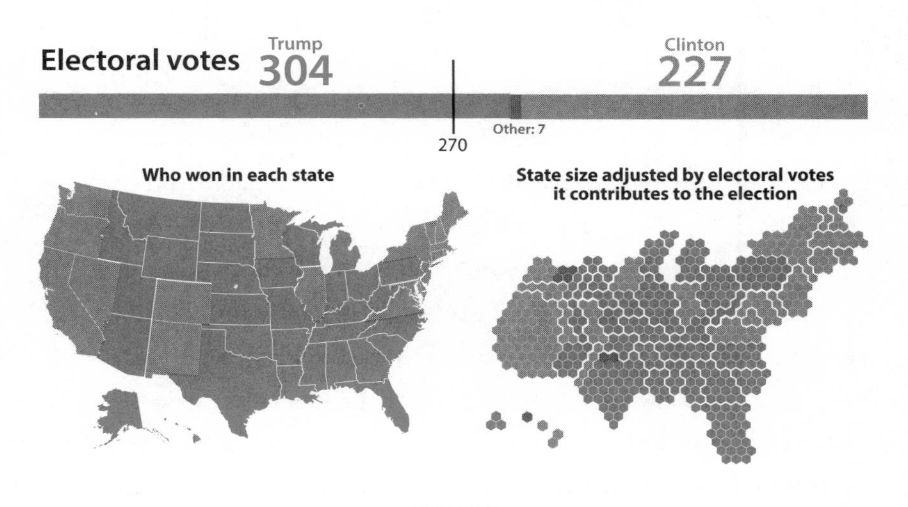

Maps are among the many kinds of charts you'll learn about in this book. Sadly, they are among the most misused. In July of 2017, I read that a popular U.S. singer called Kid Rock was planning to run for the Senate in the 2018 election.[6] He'd later claim that it was all a joke,[7] but it sounded like a serious bid at the time.

I didn't know much about Kid Rock, so I wandered through his social media accounts and saw some of the merchandise he was selling in his online store, KidRock.com. I love graphs and maps, so one T-shirt with an intriguing map of the results of the 2016 election was irresistible. Its legend indicated that, according to Mr. Rock, the results of the election matched the boundaries of two separate countries:

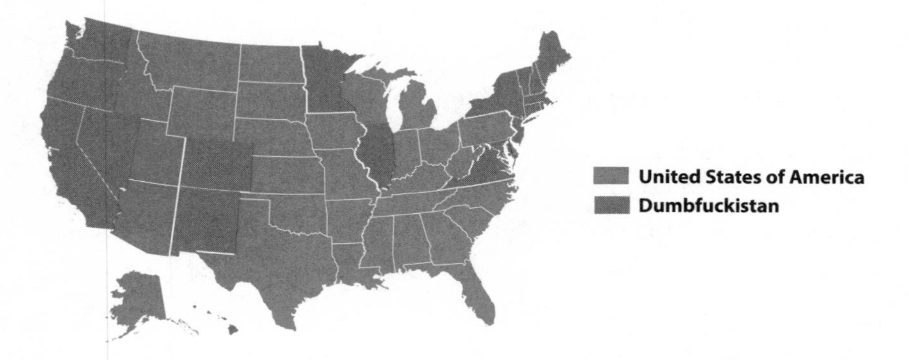

As you might now expect, this map isn't an accurate representation of the borders between the United States (read: Republican America) and Dumbfuckistan (read: Democratic America). An electoral precinct- or county-level map may be much more accurate.

Now, as an aside, I want to point out that I lived in North Carolina between 2005 and 2008. Originally from Spain, I knew little about the Tar Heel State before I arrived, other than that it was often red on the presidential election maps I'd always seen in Spanish newspapers. I was expecting to settle in a conservative place. Fine with me. I'm ideologically moderate. But my expectations were misguided. To my surprise, when I arrived, I didn't land in the United States of America—if we follow Kid Rock's nomenclature—I landed in deep Dumbfuckistan! The Chapel Hill–Carrboro area, in Orange County (North Carolina), where I lived, is quite progressive and liberal, more so than most of the rest of the state.

The city where I am now, Kendall (Florida), part of the greater Miami area, is also quite proud of its Dumbfuckistani heritage. The following maps reveal what I'd say are the true borders between the two countries Mr. Rock's T-shirt depicts:

President Donald Trump gave his first State of the Union address on January 30, 2018. Pundits on the right sang praises to his great performance as he read from a teleprompter, and those on the left criticized him. Trump devoted some time to talking about crime and got the attention of economist and Nobel Prize winner Paul Krugman, a columnist for the *New York Times*.

On several occasions during the presidential campaign in 2016, and also during his first year in office, Trump mentioned a supposedly sharp increase of violent crime in the United States, particularly murders. Trump blamed undocumented immigrants for this, an assertion that has been debunked many times over and that Krugman called a "dog whistle" in his column.[8]

However, Krugman didn't stop there. He added that Trump wasn't "exaggerating a problem, or placing the blame on the wrong people. He was inventing a problem that doesn't exist" as "there is no crime wave—there have been a few recent bobbles, but many of our big cities have seen both a surge in the foreign-born population and a dramatic, indeed almost unbelievable, decline in violent crime."

Here's a chart that Krugman provided as evidence:

It seems that what Krugman said is true: the United States has witnessed a noticeable drop in murders since the peaks in the 1970s, 1980s, and early 1990s. The trend is similar for violent crime in general.

However, isn't it odd that an article published at the beginning of 2018 includes only years up to 2014? While detailed crime statistics are hard to obtain, and it would be impossible to get a good estimate up to the day when Krugman's column was published, the FBI already had solid stats for 2016 and a preliminary estimate for 2017.[9] This is what the chart looks like if we add those years. The murder rate increased in 2015, 2016, and 2017. It doesn't look like a "bobble" at all:

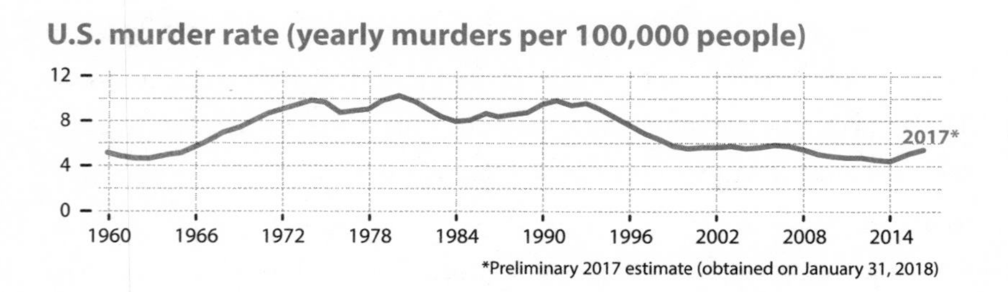

I doubt that someone with Krugman's record would conceal relevant data intentionally. Based on my own experience as a chart designer and journalist who's made plenty of silly mistakes, I've learned to never attribute to malice what could be more easily explained by absentmindedness, rashness, or sloppiness.

It's true, as Krugman wrote, that the murder rate today is way lower than it was thirty years ago. If you zoom out and take a look at the entire chart, the overall long-term trend is one of decline. Tough-on-crime politicians and pundits often ignore this, quite conveniently, and focus instead on the last few years.

However, the uptick since 2014 is relevant and shouldn't be concealed. How relevant is it, though? That depends on where you live.

This national murder rate chart, as simple and easy to read as it looks, *hides as much as it reveals.* This is a common feature of charts, since they are usually simplifications of very complex phenomena. Murders aren't increasing everywhere in the United States. Most places in the U.S. are pretty safe.

Instead, murder in the U.S. is a localized challenge: some neighborhoods in mid-sized and big cities have become so violent that they distort the national rate.[10] If we could plot those neighborhoods on the chart, they would be way above its upper gridline, perhaps even beyond the top edge of the page! If we took them off the chart, the national-murder-rate line might stay flat or even go down in recent years.

Doing this wouldn't be appropriate, of course: those cold numbers represent people being killed. However, we can and *should* demand that, when discussing data like this, politicians and pundits mention *both* overall rates *and the extreme values—also called "outliers"—that may be distorting those rates.*

Here's an analogy to convey the statistics and help you grasp the role of outliers: Imagine you're in a bar enjoying a beer. Eight other people are drinking and chatting. None of you has killed anyone in your life. Then, a tenth person comes in, a hitman for the Mob who's dispatched 50 rivals in his career. Suddenly, the average kill count per drinker in the bar jumps to 5! But of course that doesn't automatically make you an assassin.

Charts may lie, then, because they display either the wrong information or too little information. However, a chart can show the right type and amount of information and lie anyway due to poor design or labeling.

In July 2012, Fox News announced that President Barack Obama was planning to let President George W. Bush's cuts to the top federal tax rate expire by the beginning of 2013. The very wealthy would see their tax bills increase. By how much? Please estimate the height of the second bar in comparison with the first one, which represents the top tax rate under President Bush. It's a massive tax increase!

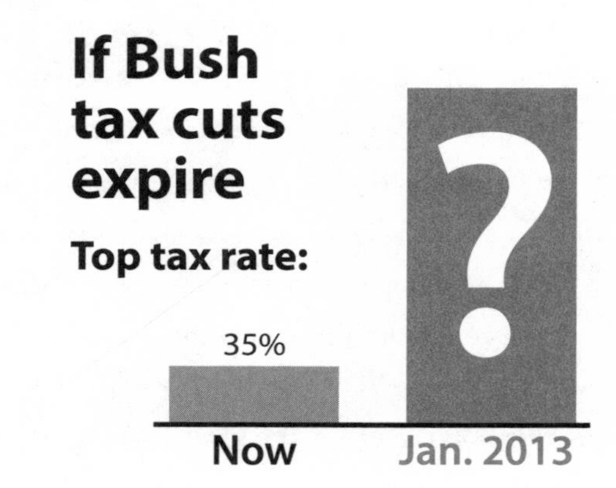

The chart that Fox displayed for a few seconds contained the figures, but they were quite tiny and hard to read. Notice that the tax increase was roughly five percentage points, but the bars were grossly misshapen to exaggerate it:

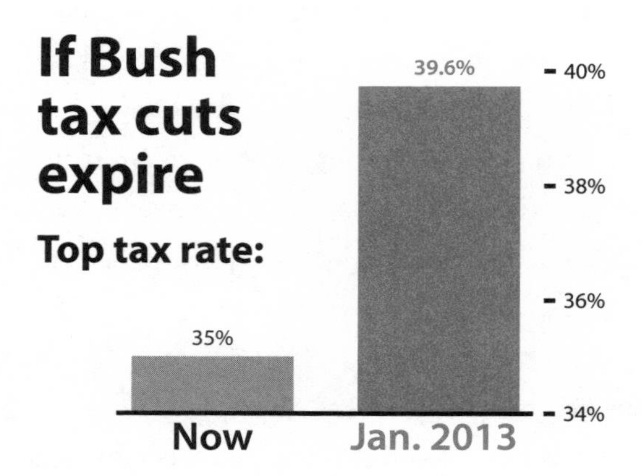

I like taxes as little as anyone else, but I dislike arguments defended with dubious charts even more, regardless of their creators' political leanings. Whoever designed this chart broke an elementary principle of chart design: if your numbers are represented by the length or height of objects—bars, in this case—the length or height should be *proportional* to those numbers. Therefore, it's advisable to put the baseline of the chart at zero:

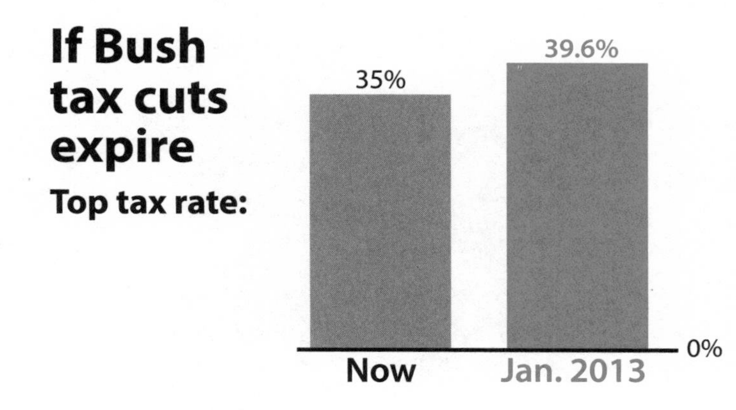

Starting a bar chart at a baseline different from zero is the most conspicuous trick in the book to distort your perception of numbers. But fudging with scales is just one of the many strategies used by swindlers and liars from all ideological denominations. There are many others that are far less easy to spot, as we'll soon see.

Even if a chart is correctly designed, it may still deceive us because we don't know how to read it correctly—we can't grasp its symbols and grammar, so to speak—or we misinterpret its meaning, or both. Contrary to what many people believe, most good charts aren't simple, pretty illustrations that can be understood easily and intuitively.

On September 10, 2015, the Pew Research Center published a survey testing U.S. citizens' knowledge of basic science.[11] One of the questions asked participants to decode the following chart. Try to read it and don't worry if you get it wrong:

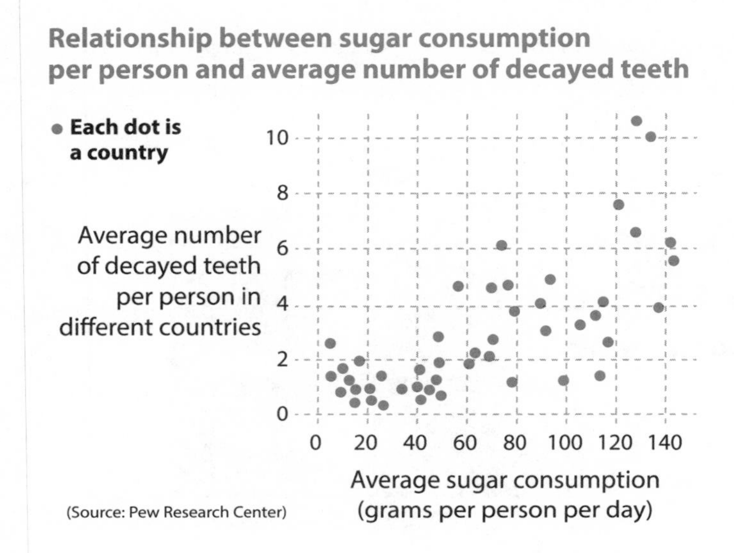

In case you've never seen a chart like this, it's called a *scatter plot*. Each dot is a country; we don't need to know which one. The position of each dot on the horizontal axis corresponds to the daily sugar consumption per person. In other words, the farther to the right a dot is, the more sugar people consume in that country, on average.

The position of a dot on the vertical axis corresponds to the number of decayed teeth per person. Therefore, the higher up a dot is, the more bad teeth people in that country have, on average.

You've probably detected a pattern: in general, and with some exceptions, the farther to the right a dot is, the higher up it tends to be as well. This is called a positive *correlation* between two metrics: sugar intake is positively correlated with worrisome dental health at the country level. (This chart on its own does not prove that more sugar leads to more decayed teeth, but we'll get to that soon.) Correlations can also be negative; for instance, the more educated countries are, the smaller the percentage of poor people they usually have.

The scatter plot is a kind of chart that is almost as old as those we all learn to read in elementary school, such as the bar graph, the line graph, and the pie chart. Still, roughly 4 out of 10 people in the survey (37%) couldn't interpret it correctly. This may have to do in part with how the questions in the survey were posed or some other factors, but it still suggests to me that a large portion of the population struggles to read charts that are commonplace in science and that are also becoming common in the news media.

And it's not just scatter plots. It also happens with charts that, at least at first glance, look easy to read. A group of researchers from Columbia University showed the following pictorial chart to more than 100 people:[12]

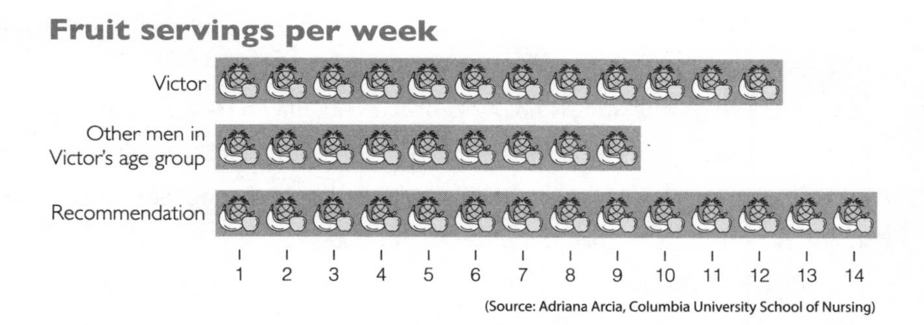

The chart reveals that "Victor," an imaginary fellow, is consuming more fruit servings per week than other men of his age, but fewer than the recommended 14 servings per week.

What the chart is intended to say is: "Victor is currently eating 12 servings *of any kind of fruit* every week. He's eating more than the average man in his age group, but 12 servings aren't enough. He should be eating 14."

Many participants read the chart too literally. They thought that Victor needed to eat the exact same amount *and kinds* of fruits portrayed in the chart 14 times every week! A participant even complained, "But a whole pineapple?" The results were similar if the icon used to represent "fruit serving" was a single apple. In that case, one participant complained about the "monotony" of eating the same fruit every day.

Charts are still seductive and persuasive, whether or not many people are able to read them correctly. In 2014, a team of researchers from New York University conducted several experiments to measure how persuasive charts are in comparison with textual information.[13] They wanted to see whether three charts—about the corporate income tax, incarceration rates, and the reasons children play video games—modified people's opinions. For instance, in the case of video games, the goal was to show participants that, contrary to some messages in the media, children don't play video games because they enjoy violence, but because they want to relax, let their imaginations fly, or socialize with friends.

Many participants' minds changed because of the charts, particularly if they didn't have strong preexisting opinions about the charts' topics. The authors conjectured that this happened "partially due to the increased sense of objectivity" that "evidence supported by numbers carries."

Studies like this have limitations, as their authors themselves acknowledged. For instance, it's hard to tell what exactly participants found persuasive: Was it the visual representation of the numbers or the numbers themselves? As the saying goes, more research is needed, but the tentative evidence we have suggests that many of us are cajoled by the mere presence of numbers and charts in the media we consume, no matter whether we can interpret them well.

The persuasiveness of charts has consequences. Very often, charts lie to us because we are prone to lying to ourselves. We humans employ numbers and charts to reinforce our opinions and prejudices, a psychological propensity called the confirmation bias.[14]

Republican congressman Steve King, a strong proponent of strict limits to immigration, posted on Twitter in February 2018:

> Illegal immigrants are doing what Americans are reluctant to do. We import young men from cultures with 16.74 times the violent death rate as the U.S. Congress has to KNOW more Americans will die as a result.[15]

King also added a table. The United States isn't shown, but it's in the 85th position, with a violent death rate of around 6 per 100,000 people:

Violent death rate per 100,000 people

Rank	Country	Rate
1	El Salvador	93
2	Guatemala	71
3	Venezuela	47
4	Trinidad-Tobago	43
5	Belize	43
6	Lesotho	42
7	Colombia	37
8	Honduras	36
9	Swaziland	36
10	Haiti	35
11	Panama	34
12	D.R. Congo	31
13	Brazil	31
14	South Africa	29
15	Mexico	27
16	Jamaica	27
17	Guyana	26
18	Rwanda	24
19	Nigeria	21
20	Uganda	20

King was fooled by his own data and chart and, as a result, he likely also fooled some of his constituents and followers. These countries are very violent, yes, but you cannot infer *from the chart alone* that the people moving from them to the United States have violent inclinations. The opposite may be true! It may well be that immigrants and refugees from dangerous countries are the meek and the peaceful, fleeing from societies where they can't work hard and thrive because they're being harassed by criminals.

To give you an anecdotal analogy, an enormous number of Spanish men my age love soccer, bullfighting, Flamenco dance, and the reggaeton song "Despacito." I'm a Spaniard, but I don't like any of those, and neither do any of my closest Spanish friends, who prefer to engage in much dorkier recreations, such as strategic board games and reading comic books, popular-science books, and science fiction. We must always be wary of inferring features of *individuals* based on statistical patterns of *populations*. Scientists call this the ecological fallacy.[16] You'll soon learn more about it.

Charts may lie in multiple ways: by displaying the wrong data, by including an inappropriate amount of data, by being badly designed—or, even if they are professionally done, they end up lying to us because we read too much into them or see in them what we want to believe. At the same time, charts—good and bad—are everywhere, and they can be very persuasive.

This combination of factors may lead to a perfect storm of misinformation and disinformation. We all need to turn into attentive and informed chart readers. We must become more *graphicate*.

Geographer William G. V. Balchin coined the term "graphicacy" in the 1950s. In a 1972 address to the annual conference of the Geographical Association, he explained its meaning. If literacy, said Balchin, is the ability to read and write, articulacy is the ability to speak well, and numeracy the ability to manipulate numerical evidence, then graphicacy is the ability to interpret visuals.[17]

The term "graphicacy" has appeared in numerous publications since then. Two decades ago, cartographer Mark Monmonier, author of the classic book *How to Lie with Maps*, wrote that any educated adult should possess a good level of not just literacy and articulacy but also numeracy and graphicacy.[18]

This is even truer now. Public debates in modern societies are driven by statistics, and by charts, which are the visual depiction of those statistics. To participate in those discussions as informed citizens, we must know how to decode—and use—them. By becoming a better chart reader, you may also become a better chart designer. Making charts isn't magic. You can create them with programs installed on common personal computers or available on the web, such as Sheets (Google), Excel (Microsoft), Numbers (Apple), open-source alternatives such as LibreOffice, and many others.[19]

By now you've seen that charts can indeed lie. I hope to prove to you, however, that by the end of this book you'll be able to not only spot the lies but also recognize the truths in good charts. Charts, if designed and interpreted properly, can indeed make us smarter and inform conversations. I invite you to open your eyes to their wondrous truths.

Chapter 1

How Charts Work

The first thing to know about charts is this:

Any chart, no matter how well designed, will mislead us if we don't pay attention to it.

What happens after we *do* pay attention to a chart, though? We need to be able to read it. Before we learn how charts lie, we must learn how they are supposed to work when they are appropriately built.

Charts—also called *visualizations*—are based on a grammar, a vocabulary of symbols, and plenty of conventions. Studying them will immunize us against many abuses.

Let's begin with the basics.

The year 1786 saw the publication of a most unusual book with what seemed to be an ill-fitting title: *The Commercial and Political Atlas*, by polymath William Playfair. "An atlas?" readers at the time may have wondered while browsing through its pages. "This book doesn't contain any map whatsoever!" But it did. One of Playfair's graphics can be seen below.

You probably have identified this chart as a common line chart, also called a *time-series line graph*. The horizontal axis is years, the vertical axis measures a magnitude, and the two lines waving through the picture represent the variation of that magnitude. The darker line (on top) is exports from

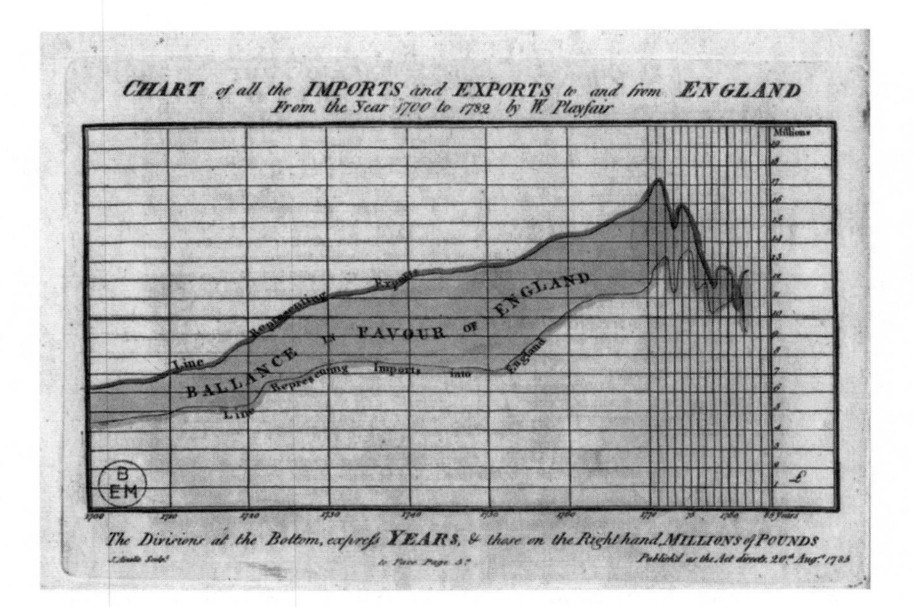

England to other countries, and the lighter line is imports into England. The shaded area in between the lines emphasizes the balance of trade, the difference between exports and imports.

Explaining how to read a chart like this may seem unnecessary nowadays. My eight-year-old daughter, now in third grade, is used to seeing them. But that wasn't the case at the end of the eighteenth century. Playfair's *Atlas* was the first book to systematically depict numbers through charts, so he devoted plenty of space to *verbalizing* what readers were seeing.

Playfair wrote his explanations because he knew that charts are rarely intuitive at a quick glance: like written language, they are based on symbols, the rules (syntax or grammar), which guide how to arrange those symbols so they can carry meaning, and the meaning itself (semantics). You can't decode a chart if you don't grasp its vocabulary or its syntax, or if you can't make the right inferences based on what you're looking at.

Playfair's book contains the word "atlas" because it truly *is* an atlas. Its maps may not represent geographic locations, but they are based on principles borrowed from traditional mapmaking and geometry.

Think of how we locate any point on the surface of the world. We do it by figuring out its coordinates, longitude and latitude. For instance, the Statue of Liberty is located 40.7 degrees north of the equator, and 74 degrees west of the Greenwich meridian. To plot its position, I simply need a map on which I can overlay a grid of subdivisions over the horizontal axis (the longitude) and vertical axis (the latitude):

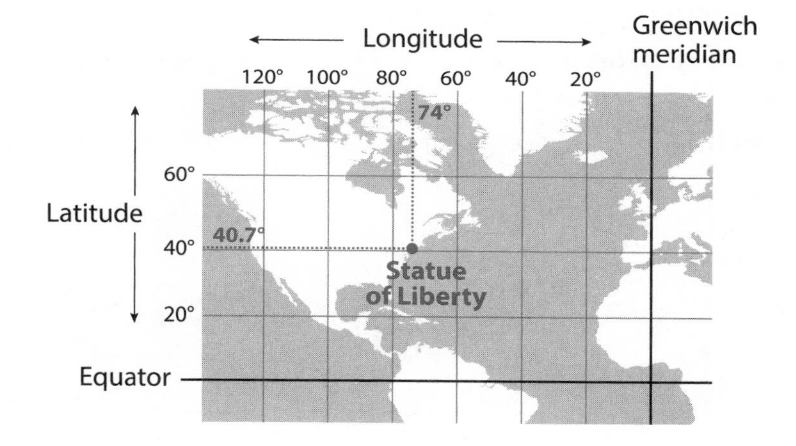

Playfair's insight, which led him to create the first line graphs and bar graphs, was this: as longitude and latitude are quantities, they can be substituted by any other quantity. For instance, year is used instead of longitude (horizontal axis), and exports/imports instead of latitude (vertical axis).

Playfair employed a couple of simple elements that lie at the core of how most charts work: the *scaffolding* of a chart and its methods of *visual encoding.*

This is where I get a bit technical, but I promise that the little extra effort this chapter requires will pay off later. Moreover, what I'll explain will prepare you for most of the charts you see anywhere and everywhere. Bear with me. Your patience will be rewarded.

To read a chart well, you must focus on the features that surround the content and support it—the chart's scaffolding—and on the content itself—how the data is represented, or *encoded.*

The scaffolding consists of features such as titles, legends, scales, bylines (who made the graphic?), sources (where did the information come from?), and so forth. It's critical to read them carefully to grasp what the chart is about, what is being measured, and how it's being measured. Here are some examples of charts with their content displayed with and without scaffolding:

Complete chart

Murders
Rate per 100,000 people (2015)

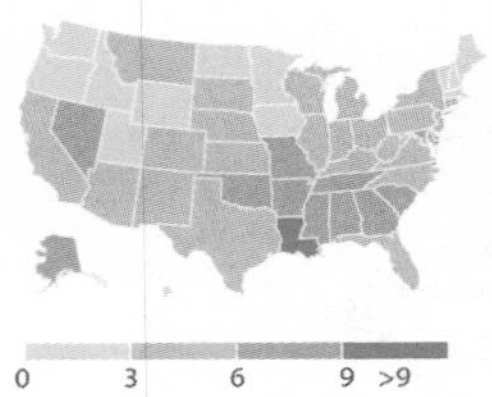

(Source: FBI Uniform Crime Reports)

Scaffolding

Murders
Rate per 100,000 people (2015)

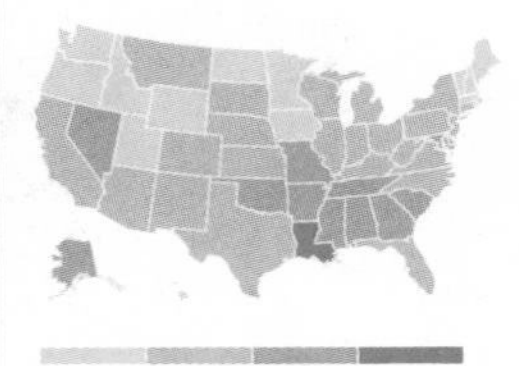

(Source: FBI Uniform Crime Reports)

Content

Murders
Rate per 100,000 people (2015)

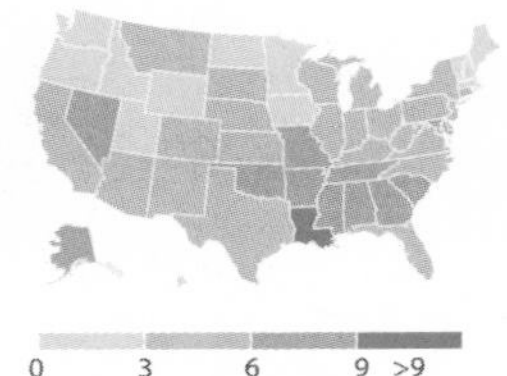

(Source: FBI Uniform Crime Reports)

Complete chart

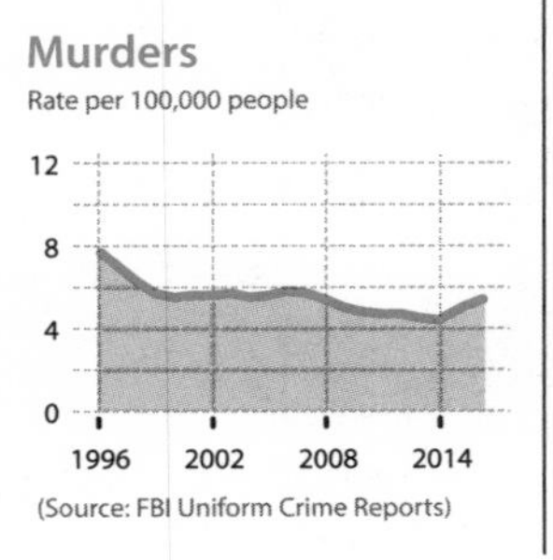

(Source: FBI Uniform Crime Reports)

Scaffolding

Murders
Rate per 100,000 people
12
8
4
0
1996 2002 2008 2014

(Source: FBI Uniform Crime Reports)

Content

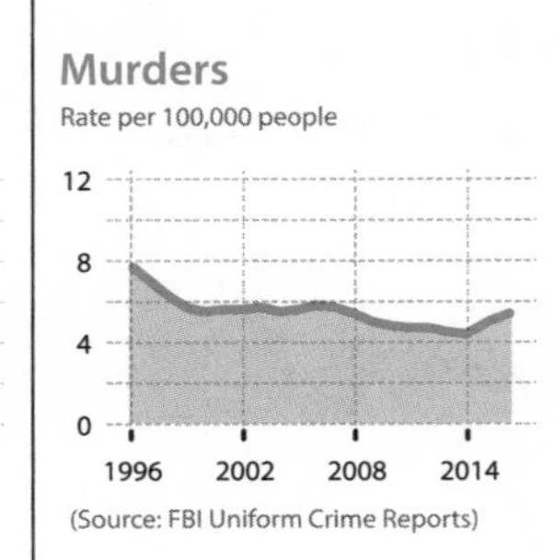

(Source: FBI Uniform Crime Reports)

The scaffolding of the map includes a legend based on sequential shades of color, which indicate a higher murder rate (darker shades) or a smaller one (lighter shades). The scaffolding of the line graph is made of a title, a subtitle indicating the unit of measurement ("rate per 100,000 people"), labels on the horizontal and vertical scales to help you compare the years, and the source of the data.

Sometimes, short textual notes may supplement a chart, emphasizing or clarifying some important points. (Imagine that I had added the explainer "Louisiana has the highest murder rate in the United States, 11.8 per 100,000 people.") We call this the "annotation layer," a term coined by designers who work for the graphics desk of the *New York Times*. The annotation layer is also part of a chart's content.

The core element of most charts is their *visual encodings*. Charts are always built with symbols—usually, but not always, geometric: rectangles, circles, and the like—that have some of their properties varied according to the numbers they represent. The property that we choose to change depending on our data is the encoding.

Think of a bar graph. In it, bar length or height varies in proportion to underlying numbers; the bigger the number, the longer—or taller—the bar will be:

Compare India and the United States. The population of India is roughly four times the size of the U.S. population. Therefore, as our chosen method of encoding is *length*, India's bar must be four times the length of that of the United States.

There are many other encodings that can be used in charts in addition to length or height. One very popular method is *position*. In the chart below, the position of each county in Florida—symbolized by a dot—on the horizontal (x) axis corresponds to the yearly income per person. The farther a dot is to the right, the richer the typical individual is in that county:

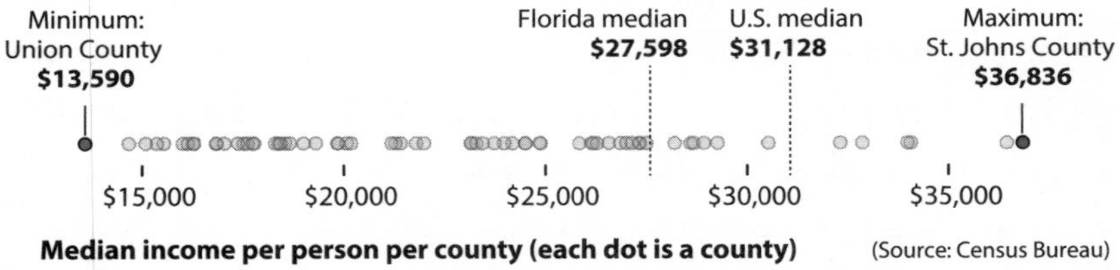

Minimum:
Union County
$13,590

Florida median
$27,598

U.S. median
$31,128

Maximum:
St. Johns County
$36,836

$15,000 $20,000 $25,000 $30,000 $35,000

Median income per person per county (each dot is a county) (Source: Census Bureau)

This chart compares the median income in each Florida county. The median is the score that splits the full range of values into two halves of equal size. Here's an example: Union County has a median income of $13,590. Its population is roughly 15,000. Therefore, what the median tells us is that around 7,500 people living in Union County make more than $13,590 every year and the other 7,500 make less—but we don't know *how much* more or less: some people may have an income of $0, and others may make millions.

Why are we using the median and not the more widely known arithmetic mean, also known as the average? We do this because the mean is very sensitive to extreme values and, as such, is much higher than typical incomes. Imagine this situation: You want to study the incomes of a county with 100 inhabitants. Ninety-nine of them have a yearly income very close to $13,590. But one of them makes $1 million a year.

The median of this distribution would still be $13,590: half the people are a bit poorer than that, and the other half—the half that includes our very wealthy friend—are richer. But the mean would be much higher: $23,454. This is the result of adding up the incomes of all inhabitants in the county and dividing the result by 100 people. As the old saying goes, whenever Bill Gates participates in a meeting, everyone in the room becomes a millionaire, if we take the arithmetic mean of the group's wealth.

Let's return to my dot chart. A large chunk of our brains is devoted to processing information that our eyes gather. That is why it's often easier to spot interesting features of numbers when those numbers are represented through visual encodings. Take a look at the numerical table—a kind of chart, but not one that uses visual encodings—with all Florida counties and their corresponding median incomes.

County	Income per person ($)	County	Income per person ($)	County	Income per person ($)
Alachua County	24,857	Hamilton County	16,295	Nassau County	28,926
Baker County	19,852	Hardee County	15,366	Okaloosa County	28,600
Bay County	24,498	Hendry County	16,133	Okeechobee County	17,787
Bradford County	17,749	Hernando County	21,411	Orange County	24,877
Brevard County	27,009	Highlands County	20,072	Osceola County	19,007
Broward County	28,205	Hillsborough County	27,149	Palm Beach County	32,858
Calhoun County	14,675	Holmes County	16,845	Pasco County	23,736
Charlotte County	26,286	Indian River County	30,532	Pinellas County	29,262
Citrus County	23,148	Jackson County	17,525	Polk County	21,285
Clay County	26,577	Jefferson County	21,184	Putnam County	18,377
Collier County	36,439	Lafayette County	18,660	St. Johns County	36,836
Columbia County	19,306	Lake County	24,183	St. Lucie County	23,285
DeSoto County	15,088	Lee County	27,348	Santa Rosa County	26,861
Dixie County	16,851	Leon County	26,196	Sarasota County	32,313
Duval County	26,143	Levy County	18,304	Seminole County	28,675
Escambia County	23,441	Liberty County	16,266	Sumter County	27,504
Flagler County	24,497	Madison County	15,538	Suwannee County	18,431
Franklin County	19,843	Manatee County	27,322	Taylor County	17,045
Gadsden County	17,615	Marion County	21,992	Union County	13,590
Gilchrist County	20,180	Martin County	34,057	Volusia County	23,973
Glades County	16,011	Miami-Dade County	23,174	Wakulla County	21,797
Gulf County	18,546	Monroe County	33,974	Walton County	25,845
Florida median	27,598			Washington County	17,385
U.S. median	31,128				

Tables are great when we want to identify specific individual figures, like the median income of one or two counties, but not when what we need is a bird's-eye view of all those counties together.

To see my point, notice how much easier it is to spot the following features of the data by looking at the dot chart rather than at the numerical table:

- The minimum and maximum values can be seen in comparison to the rest.
- Most counties in Florida have a lower median income than the rest of the United States.
- There are two counties, St. Johns and another one that I didn't label, where the median income is clearly much higher than in the rest of Florida.

- There is one county, Union, that is much poorer than the other poor counties in Florida. Notice that there's a gap between the position of Union County's circle and the rest of the circles.
- There are many more counties with low median incomes than counties with high incomes.
- There are many more counties below the state's median income ($27,598) than above it.

How is this last point possible? After all, I've just told you that the median is the value that splits the population in half. If that's true, half the counties on my chart should be poorer than the state median, and the other half should be richer, shouldn't they?

But that's not how this works. That number of $27,598 is not the median of the medians of the 67 Florida counties. It's the median income of the more than 20 million Floridians, *regardless of the counties where they live.* Therefore, it's half the *people* in Florida (not half the *counties*) who make less than $27,598 a year, while the other half make more.

This apparent distortion on the chart might happen because of population size differences between counties that tend to be richer and counties that tend to be poorer.

To find out, let's create a chart that still uses position as its method of encoding. See it below. The position on the *x* axis still corresponds to the median income in each county; the position on the *y* axis corresponds to the population. The resulting scatter plot indicates that my hunch may not be misguided: the median income in the most populous county in Florida, Miami-Dade, is slightly lower than the state median (its dot is to the left of the vertical red line that marks the median income at the state level). Some other big counties, such as Broward or Palm Beach, which I highlighted, have incomes above the state median.

Look closely at the many counties on the left. Individually, many are sparsely populated (position on the vertical axis), but their combined pop-

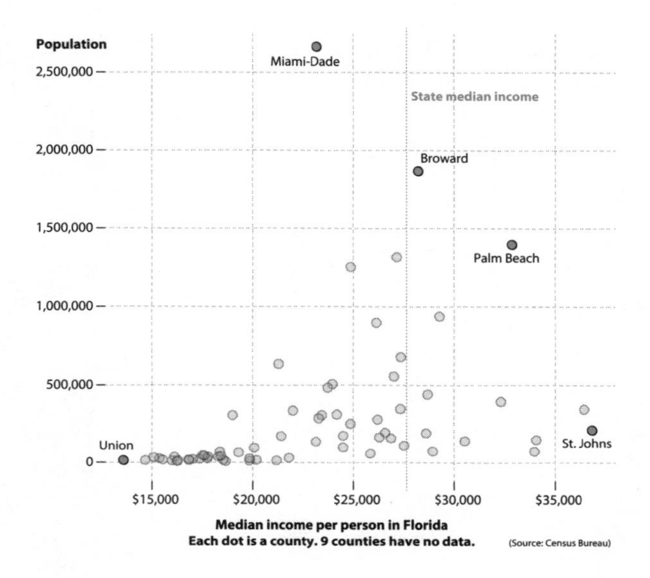

ulation balances out that of the richer counties on the right side of the chart.

We've discovered plenty of interesting features of charts just by playing with a few numbers. Let's now try something else. First, let's change the vertical axis. Instead of making the vertical position of each dot correspond to population, we'll make it correspond to the percentage of people per county who had a college degree by 2014. The higher up on the scale a county is, the larger the share of people who had a college education.

Second, let's change the size of the dots according to population density, the number of people per square mile. After length/height and position, we'll learn about another method of encoding: *area*. The bigger the bubble, the bigger the population density of the county that bubble represents. Spend some time reading the chart—again, paying attention to the position of all the dots on the horizontal and vertical scales—and think about what it reveals:

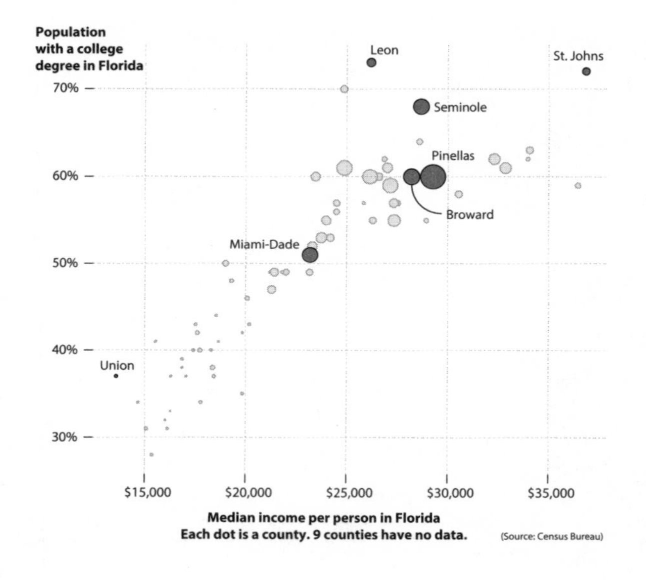

Here are some things I can perceive at a quick glance:

—In general, the higher the median income in a county (position on the horizontal axis), the higher the number of people who have a college education (position on the vertical axis). Income and education are positively associated.

—There are some exceptions to this pattern. For instance, Leon County, where the Florida state capital of Tallahassee is, has a very large percentage of college-educated people, but its median income isn't that high. This may be due to many factors. For instance, it might be that Tallahassee has large pockets of poverty but also attracts very wealthy and highly educated individuals who want to work in government or to live close to the halls of power.

—Encoding population density with the area of the bubbles reveals that counties that are richer and have many college graduates tend to be more densely populated than those that are poorer.

In case you seldom read charts, you may be wondering how it's possible to see so much so quickly. Reading charts is akin to reading text: the more you practice, the faster you'll be able to extract insights from them.

That being said, there are several tricks that we can all use. First, always peek at scale labels, to determine what it is that the chart is measuring. Second, scatter plots have that name for a reason: they are intended to show the relative *scattering* of the dots, their dispersion or concentration in different regions of the chart. The dots on our chart are quite dispersed on both the horizontal and vertical scales, indicating that median county incomes vary a lot—there are very small and very large incomes—and the same applies to college education.

The third trick is to superimpose imaginary quadrants on the chart and then name them. If you do that, even if it's just in your mind, you'll immediately grasp that there aren't any counties in the lower-right quadrant and few on the upper left. Most counties are in the top right (high income, high education) or the bottom left (low income, low education). You can see the result here:

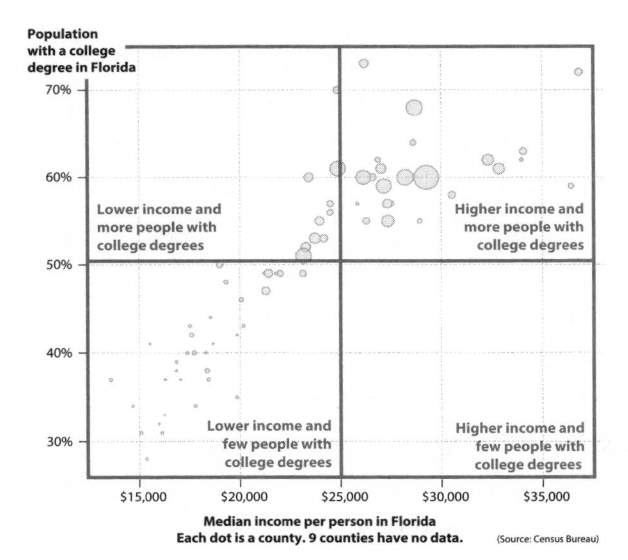

A fourth trick is to trace an imaginary line that goes roughly through the center of the bubble cloud, which reveals the overall direction of the relationship between average income per person and percentage of people who have a college education. In this case, it's a line pointing upward (I've erased the scale labels for clarity):[2]

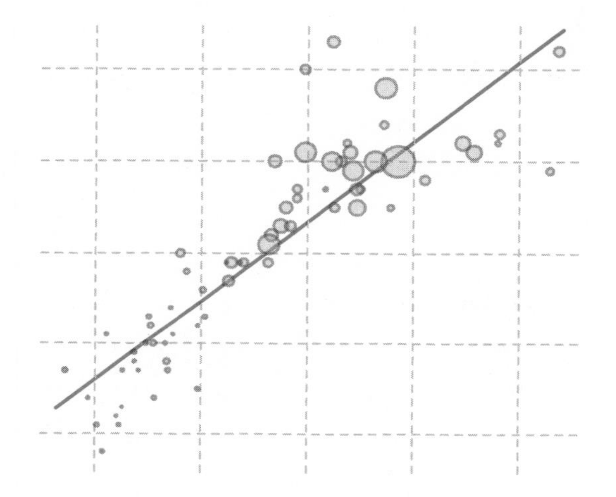

Once you apply these tricks, you'll notice that the overall direction is right and up, meaning that the more you have of the metric on the horizontal axis (income), the more you have of the metric on the vertical one (college degrees). This is a positive association. Some associations are negative, as we saw in the introduction. For instance, income is negatively correlated with poverty rates. If we put poverty rates on the vertical (y) axis, our trend curve would go down, indicating that the higher the median income of a county, the lower its poverty rate will tend to be.

We should never infer from a chart like this that the association is *causal*. Statisticians like to repeat the mantra "Correlation doesn't equal causation," although associations are often the first step toward identifying causal connections between phenomena, given that you do enough inquiries. (I'll have more to say about this in chapter 6.)

What statisticians mean is that we cannot claim *from a chart alone* that getting more college degrees leads to higher income, or vice versa. Those assertions may be right or wrong, or there may be other explanations for the high variation in median incomes and college education that our chart reveals. We just don't know. Charts, when presented on their own, seldom offer definitive answers. They just help us discover intriguing features that may later lead us to look for those answers by other means. Good charts empower us to pose good questions.

Using area as a method of encoding is quite common in maps. In the introduction we saw several bubble maps displaying the number of votes received by each major candidate in the 2016 presidential race. Here's another, with bubble area being proportional to population per county:

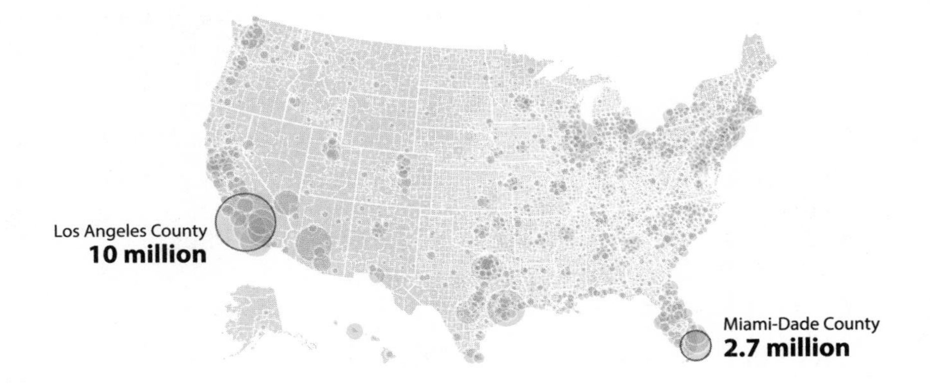

I highlighted Miami-Dade County because it's where I live, and Los Angeles County because I didn't know it was so massive. Los Angeles is the most populous county in the United States. It has nearly four times as many people as Miami-Dade. Let's put the two bubbles side by side, and also encode the data with length, on a bar chart:

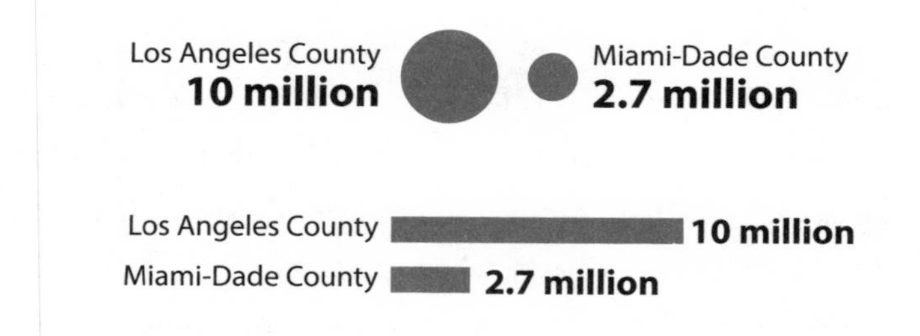

Notice that the difference between the sizes of the two counties looks less dramatic when population is encoded with *area* (bubble chart) than when it is encoded with *length* or *height* (bar chart).

Why does this happen? Think about it this way: A county of 10 million people has roughly four times the population of a county of 2.7 million people. If the objects we're using are truly proportional to the numbers they are representing, we should be able to fit four bubbles the size of Miami-Dade's into the bubble of Los Angeles, and four bars the length of Miami-Dade's into Los Angeles's bar. See for yourself (the smaller, black-outlined circles overlap, but their overlaps are similar in size to the empty spaces between them):

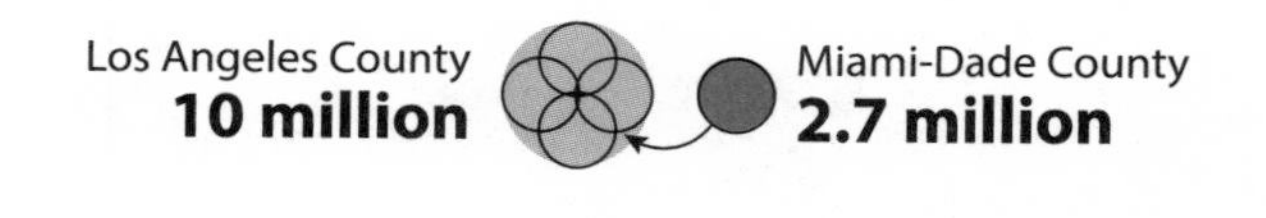

Los Angeles County 10 million
Miami-Dade County 2.7 million

A very common mistake designers make when picking bubbles to represent data is to vary not their *area* but rather their height *or* length—their *diameter*—as if they were bar charts. It's also a common trick used by those who want to exaggerate differences between numbers, so be warned.

Los Angeles has nearly 4 times the population of Miami-Dade, but if you quadruple the height of a circle, you're also quadrupling its length. Thus, you aren't making it 4 times its original size, but *16* times as big as it was before! See what happens if we scale the bubbles representing Los Angeles and Miami-

Dade according to their diameter (wrong) and not their area (right). Now we can fit 16 bubbles the size of Miami-Dade's inside Los Angeles's bubble:

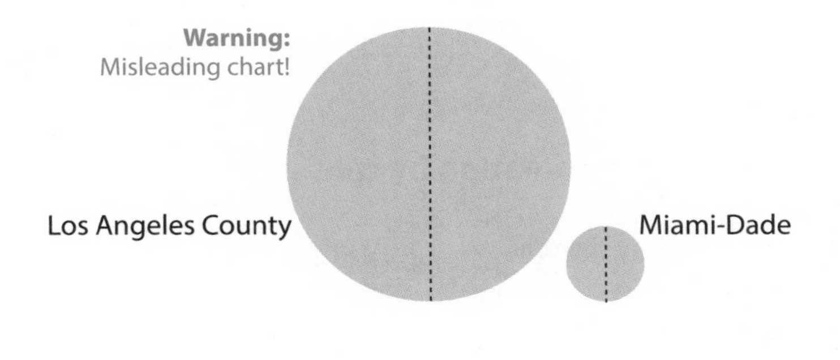

Many other kinds of charts use area as their method of encoding. For example, the treemap is increasingly favored in the news media. Paradoxically, it doesn't look like a tree at all; instead, it looks like a puzzle made of rectangles of varying sizes. Here's an example:

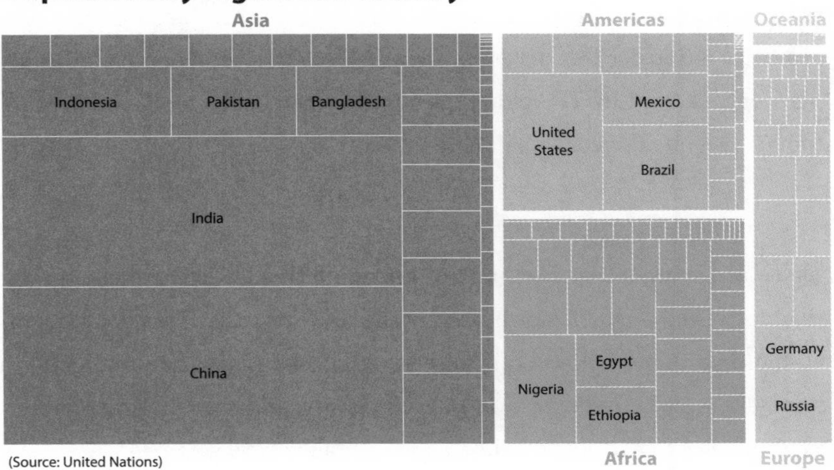

Treemaps receive their name because they present nested hierarchies.[3] In my chart, each rectangle's area is proportional to the population of a country. The combined area of the rectangles in a continent is also proportional to the population of that continent.

The treemap is sometimes used as a substitute for a more familiar graphic also based on area: the pie chart. Here's one with the same continent population data above:

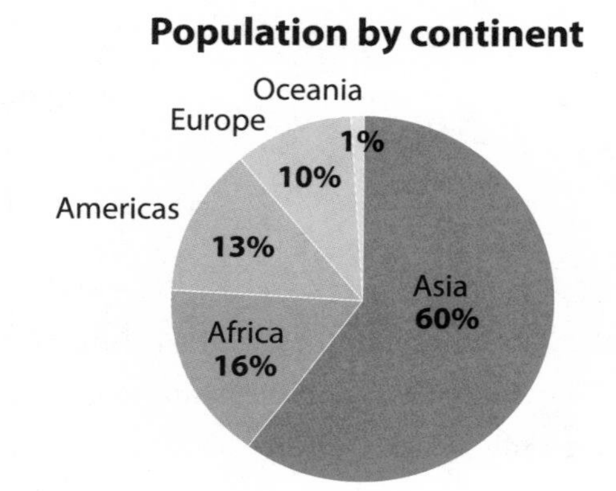

The area of the segments in a pie chart is proportional to the data, but so are its angles (*angle* is another method of encoding) and the arcs of each segment along the circumference. Here's how it works: The circumference of a circle is 360 degrees. Asia represents 60% of the world population. Sixty percent of 360 is 216. Therefore, the angle formed by the two boundaries of the Asia segment needs to be 216 degrees.

There are many other methods of encoding besides length/height, position, area, and angle. Color is among the most popular. This book begins with a map that employs both *color hue* and *color shade*: hue (red/gray) to represent which candidate won in each county, and shade (lighter/darker) to represent the winning candidate's percentage of the vote.

These two maps show the percentage of African Americans and Hispanics per U.S. county. The darker the shade of grey, the larger the percentage of the population in these counties that is either African American or Hispanic:

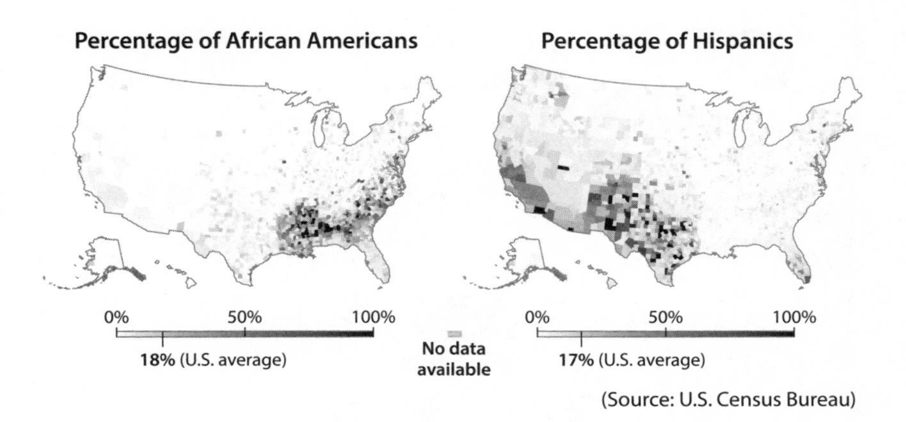

Color shades are sometimes used to great effect in a kind of chart we can call a *table heat map*. In the following chart, the intensity of red is proportional to the variation of global temperatures in degrees Celsius per month and year in comparison to the average temperature of the 1951–1980 period:

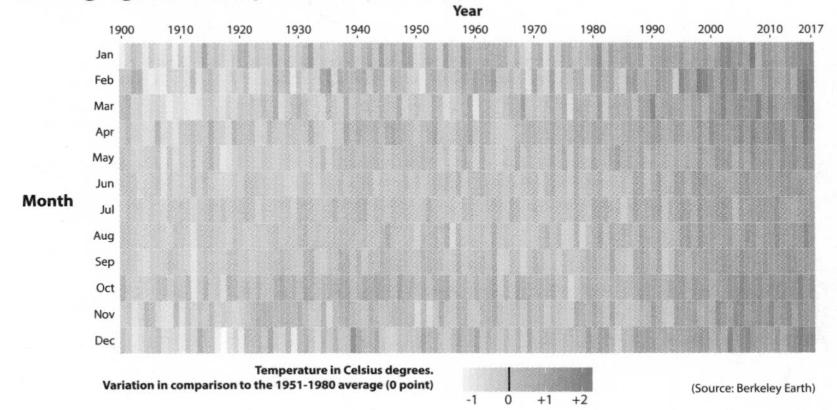

Each column of rectangles is a year, and each row is a month. The scale of a heat map isn't as precise and detailed as others we've seen because the goal of this chart isn't to focus on the specifics but on the overall change: the closer we get to the present, the hotter most months have become.

There are other ways of encoding data that are a bit less common. For instance, instead of varying the position, length, or height of objects, we could change their *width* or *thickness*, as on this chart, designed by Lázaro Gamio for the website Axios. Line width is proportional to the number of people or organizations that President Trump criticized on social media between January 20 and October 11, 2017:[4]

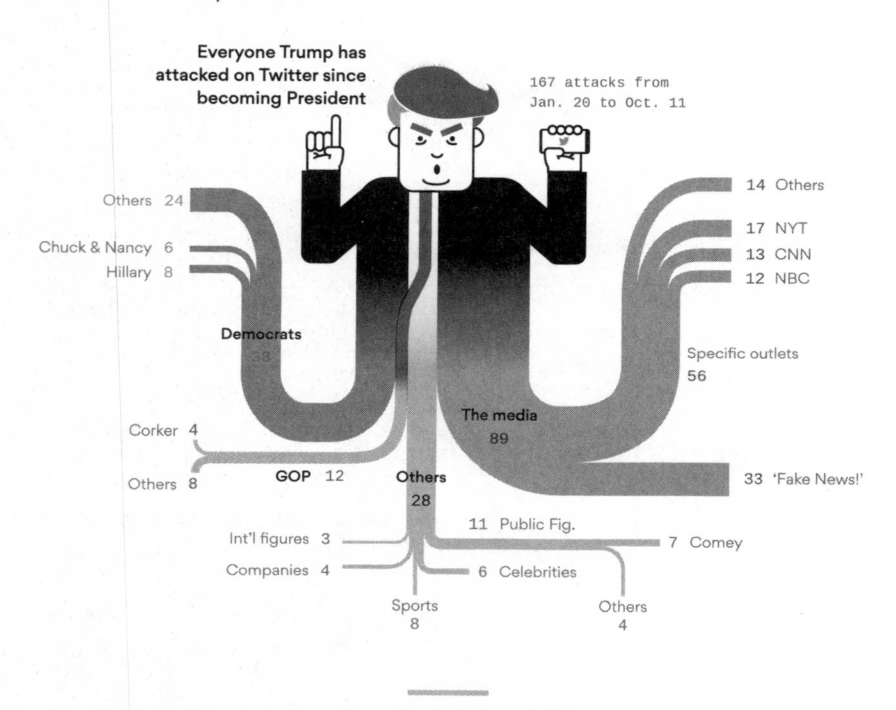

To summarize, most charts encode data through the variation of properties of symbols such as lines, rectangles, or circles. These properties are the methods of encoding we've just learned: the *height* or *length*, the *position*, the *size* or *area*, the *angle*, the *color* hue or *shade*, and so forth. We also learned that a chart can combine multiple methods of encoding.

Now, let me put you to the test. The following chart represents the fertility rates of Spain and Sweden between 1950 and 2005. Fertility rate is the average number of children per woman in a country. As you can see, in the

1950s, Spanish women had, on average, more children than did Swedish women, but the situation reversed in the 1980s. Try to identify what method or methods of encoding are at work here:

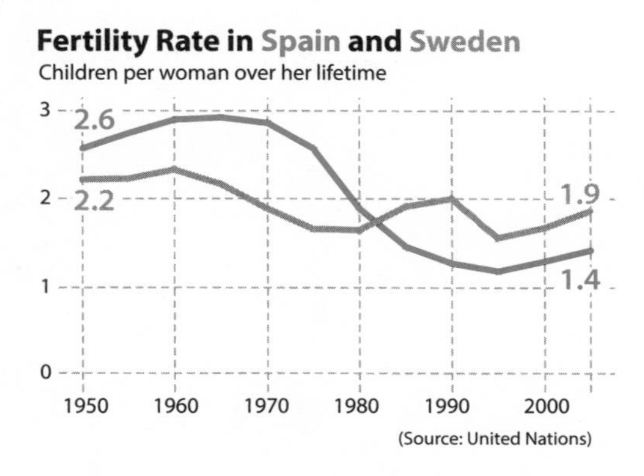

The first one is *color hue*, which is used to identify the two countries, Spain (red) and Sweden (grey).

The quantity itself, number of children per woman, is primarily encoded through *position*. Line graphs are created by placing dots on the horizontal axis—years, in this case—and on the vertical axis, in correspondence to the magnitude we're measuring, and then connecting them with lines. If I eliminate the lines, the result is still a graphic showing the variation of the fertility rates in Spain and Sweden, although it's a bit less clear:

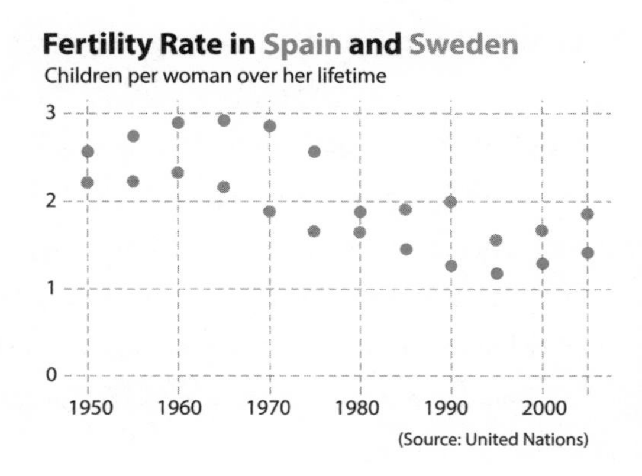

The *slope* also conveys information in a line graph, because when we connect the dots with lines, their slopes are a good indication of how steep or flat the variation is.

What about this chart? What are the methods of encoding here?

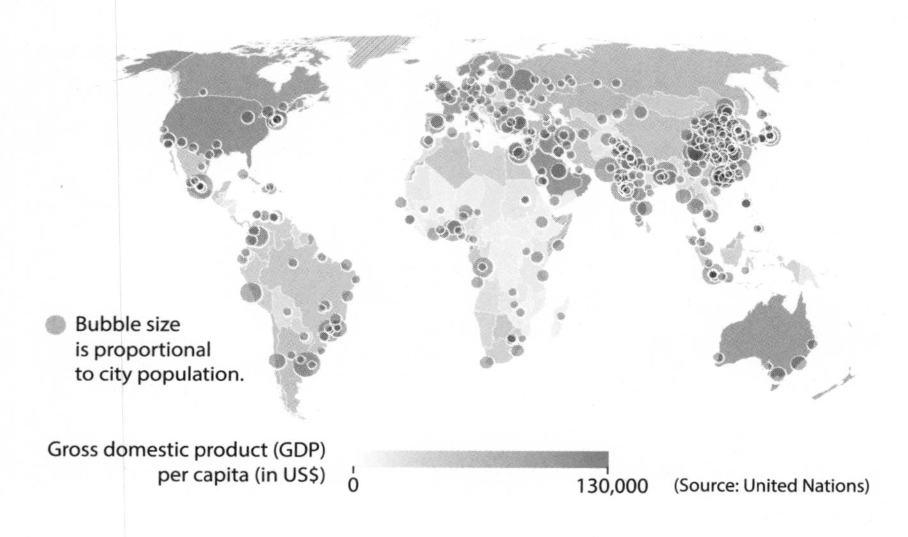

The first one you probably noticed is *color shade*: the darker the color of a country, the higher its gross domestic product per person. The second is *area*: those bubbles represent the populations of metropolitan regions that have more than one million people. That's why, for instance, Miami isn't on the map; the greater Miami area is a huge urban sprawl comprising several cities, none of them with more than one million inhabitants.

But there's more. *Position* is also a method of encoding. Why? Think of what we learned at the beginning of this chapter: maps are constructed by locating points on a plane based on a horizontal scale (longitude) and a vertical scale (latitude). The landmasses and country borders on the map are made of many little dots connected to each other, and the positions of the city bubbles are also determined by the longitude and latitude of each city.

Cognitive psychologists who have written about how we read charts point out that our prior knowledge and expectations play a crucial role. They suggest that our brains store ideal "mental models" to which we compare the graphics we see. Psychologist Stephen Kosslyn has even come up with a "principle of appropriate knowledge"[5] which, if applied to charts, means that effective communication between a designer (me) and an audience (you) requires that we share an understanding of what the chart is about and of the way data is encoded or symbolized in the chart. This means that we share roughly similar mental models of what to expect from that particular kind of chart.

Mental models save us a lot of time and effort. Imagine that your mental model of a line graph is this: "Time (days, months, years) is plotted on the horizontal axis, the amount is plotted on the vertical axis, and the data is represented through a line." If that's your mental model, you'll be able to quickly decode a chart like this without paying much attention to its axis labels or title:

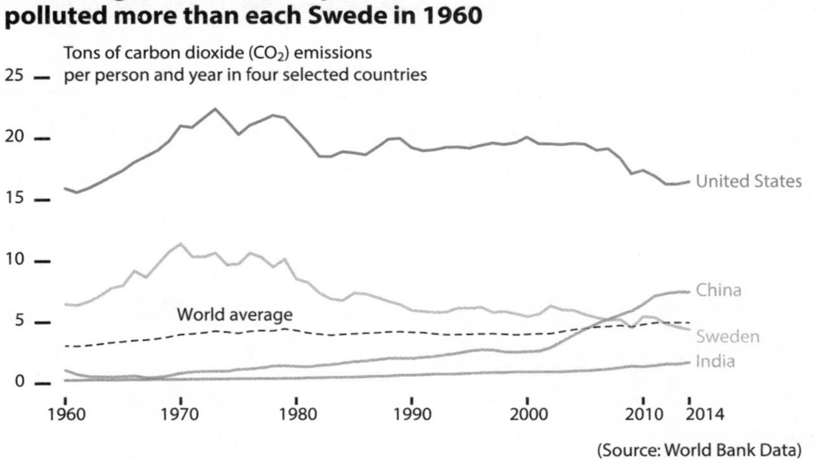

However, mental models can lead us astray. My own mental model of a line chart is much wider and more flexible than the one I described above.

If the only mental model you have for a line chart is "time on the horizontal axis, magnitude on the vertical axis," you'll likely be confused by this graphic:

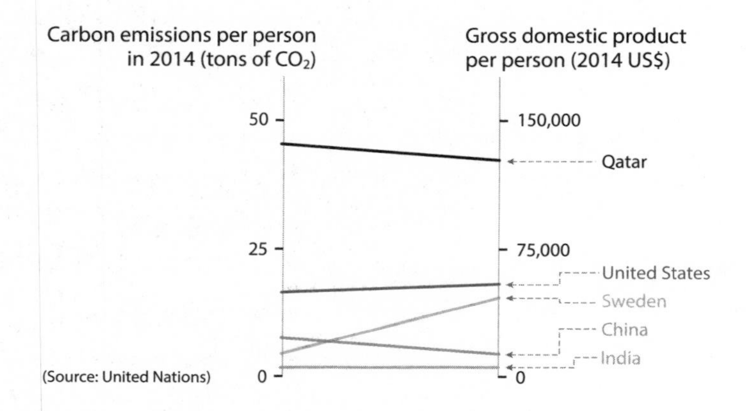

This is called a *parallel coordinates plot*. It's also a chart that uses lines, but it doesn't have time on the horizontal axis. Read the axis headers and you'll see that there are two separate variables: carbon emissions per person; and gross domestic product, or GDP, per capita in U.S. dollars. The methods of encoding here, as in all charts that use lines to represent data, are position and slope: the higher up a country is on either scale, the bigger its carbon emissions or the wealth of its people.

Parallel coordinates plots were invented to compare different variables and see relationships between them. Focus on each country and on whether its line goes up or down. The lines of Qatar, the United States, and India are nearly flat, indicating that their position on one axis corresponds to the position on the other axis (high emissions are associated with high wealth).

Now focus on Sweden: people in Sweden contaminate relatively little, but their average per capita GDP is almost as high as that of U.S. citizens. Next, compare China and India: their GDPs per capita are much closer than their CO_2 emissions per person. Why? I don't know.[6] A chart can't always answer a question, but it's an efficient way to discover intriguing facts that might ignite curiosity and cause you to pose *better* questions about data.

Here is another challenge. Now that you're well into this chapter, you'll have a pretty good mental model of a scatter plot. This one, in which I labeled some countries that I found curious, should be rather simple:

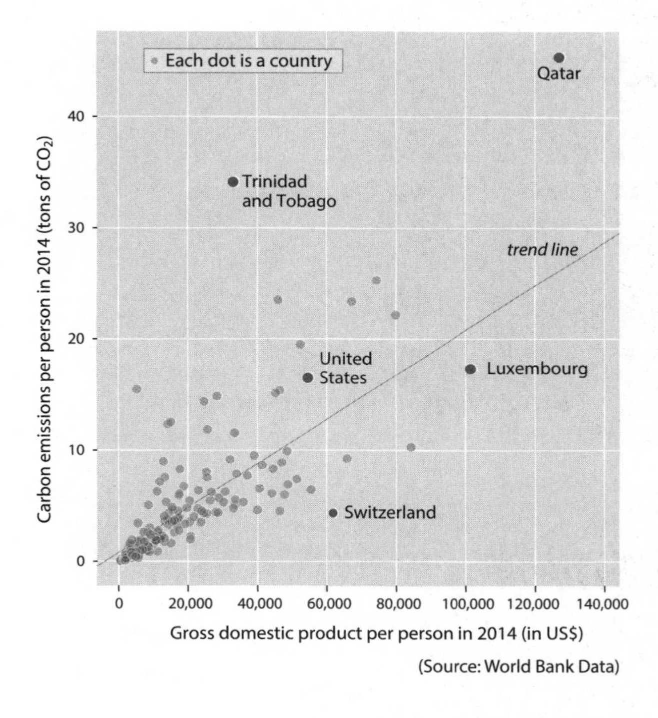

(Source: World Bank Data)

The mental model you have already developed for traditional scatter plots allows you to see that, with just some exceptions—some highlighted on the chart itself—the wealthier people in these countries are, the more they contaminate. But what if I show you another scatter plot, which looks like a line graph? You can see it on the next page.

Before your head explodes or you throw this book out the window, I should disclose that the first time I saw a chart like that, I was as befuddled as you. This type of chart, often called a *connected scatter plot*, is a bit hard to parse. Think about it this way:

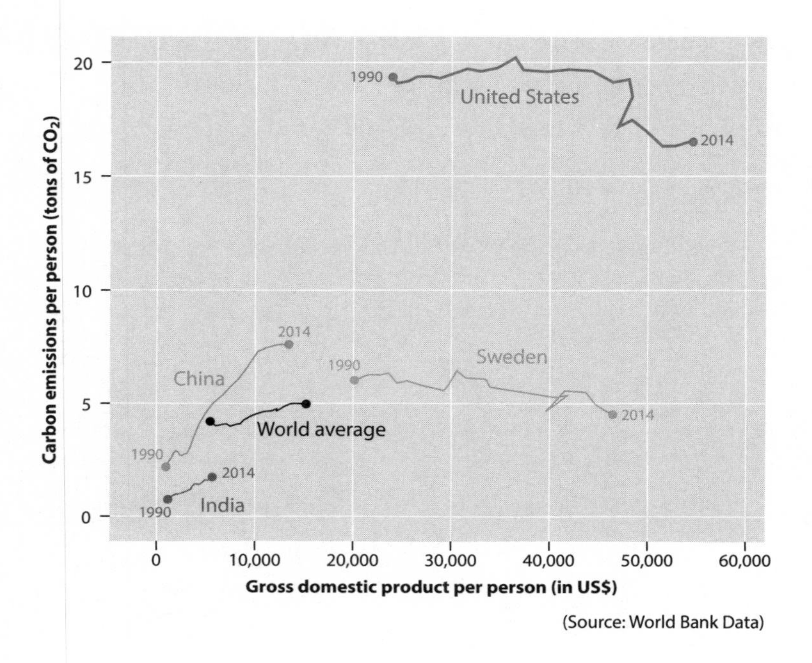

- Each line is a country. There are four country lines, plus a line for the world average.
- The lines are made by connecting dots, each corresponding to a year. I've highlighted and labeled only the first dot and the last dot on each line, corresponding to 1990 and 2014.
- The position of each dot on the horizontal axis is proportional to the GDP of the people in that country *in that year.*
- The position of each dot on the vertical axis is proportional to the amount of carbon emissions each person in those countries generated, on average, *in that year.*

The lines on this chart are like paths: they move forward or backward depending on whether people became richer or poorer, year by year, and they move up or down depending on whether people in those countries contaminated more or less. To make things clearer, let me add directional arrows to a couple of lines, plus a wind rose:

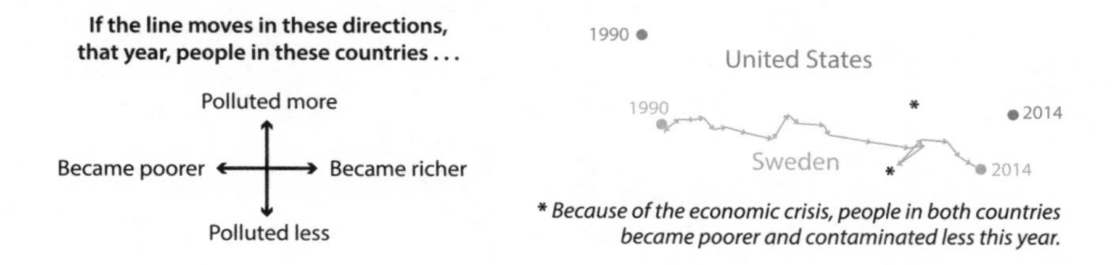

Why would anyone plot the data in such an odd way? Because of the point this chart is trying to make: at least in advanced economies, an increase in wealth doesn't always lead to an increase in pollution. In the two rich countries I chose, the United States and Sweden, people became wealthier on average between 1990 and 2014—the horizontal distance between the two points is very wide—but they also contaminated less, such that the 1990 point is higher up than the 2014 point in both cases.

The relationship between GDP and pollution is often different in developing nations, because these countries usually have large industrial and agricultural sectors that contaminate more. In the two cases I chose, China and India, people have become wealthier—the 2014 point is farther to the right than the 1990 point—and, in parallel, they also contaminate much more. As you can see if you go back to the previous chart, the 2014 point is much higher than the 1990 one.

You may be thinking—and I'd agree—that to convey this message, it would also be appropriate to show the two variables, CO_2 emissions and GDP per capita, as paired line graphs like the ones below.

Charts like the connected scatter plot are the reason why I began this chapter with the reminder that charts are seldom intuitive or self-explanatory, as many people believe. To read a chart correctly—or to create a good mental model of a type of chart we've never seen before—we must pay attention to

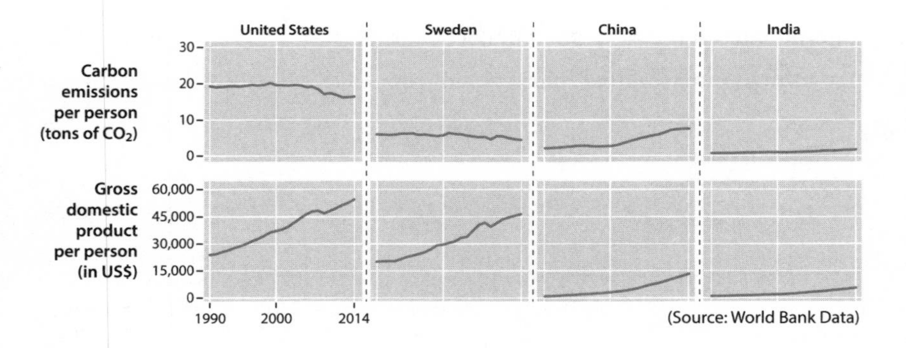

it and never take anything for granted. Charts are based on a grammar and a vocabulary made of symbols (lines, circles, bars), visual encodings (length, position, area, color, and so on), and text (the annotation layer). This makes charting as flexible as using written language, if not more.

To explain something in writing, we assemble words into sentences, sentences into paragraphs, paragraphs into sections or chapters, and so on. The order of words in a sentence depends on a series of syntactic rules, but it may vary depending on what we want to communicate and on the emotional effect we want to instill. This is the opening of Gabriel García Márquez's masterpiece, *One Hundred Years of Solitude*:

> Many years later, as he faced the firing squad, Colonel Aureliano Buendía was to remember that distant afternoon when his father took him to discover ice.

I could convey the same information by assembling the words in a different manner:

> Colonel Aureliano Buendía remembered the distant afternoon when his father took him to discover ice while facing the firing squad many years later.

The former opening has a certain musicality, while the latter is clunky and undistinguished. But both convey the same amount of information because they follow the same rules. If we read them slowly and carefully, we'll grasp their content equally well, although we'll enjoy the former more than the latter. Something similar happens with charts: if you just skim them, you won't understand them—although you may *believe* you do—and well-designed charts aren't just informative but also graceful and, like a good turn of phrase, sometimes even playful and surprising.

The same way that a long, deep, and complex sentence like that above can't be understood in the blink of an eye, charts that display rich and worthwhile information will often demand some work from you. A good chart isn't an illustration but a *visual argument* or part of an argument. How should you follow it? By taking the steps inside the red boxes I added to the following complex but illuminating chart by *Washington Post* data reporter David Byler.

1. Title, introduction (or caption), and source

If the chart includes a title and a description, read them first. If it mentions its source, take a look at it (more about this in chapter 3).

2. Measurements, units, scales, and legends

A chart must tell you what is being measured and how. Designers can do this textually or visually. Here, the vertical scale corresponds to the difference between the results of several special elections and the results of the 2016 U.S. presidential election. The horizontal scale is the number of days since Inauguration Day, January 20, 2017. The color legend tells us that the circles indicate who won each special election.

Democrats are winning special elections 1

Since Inauguration Day, in January 2017, Democratic candidates have made big gains in special elections. Numerous districts experienced a large increase of democratic votes in comparison to the results of the 2016 presidential election.

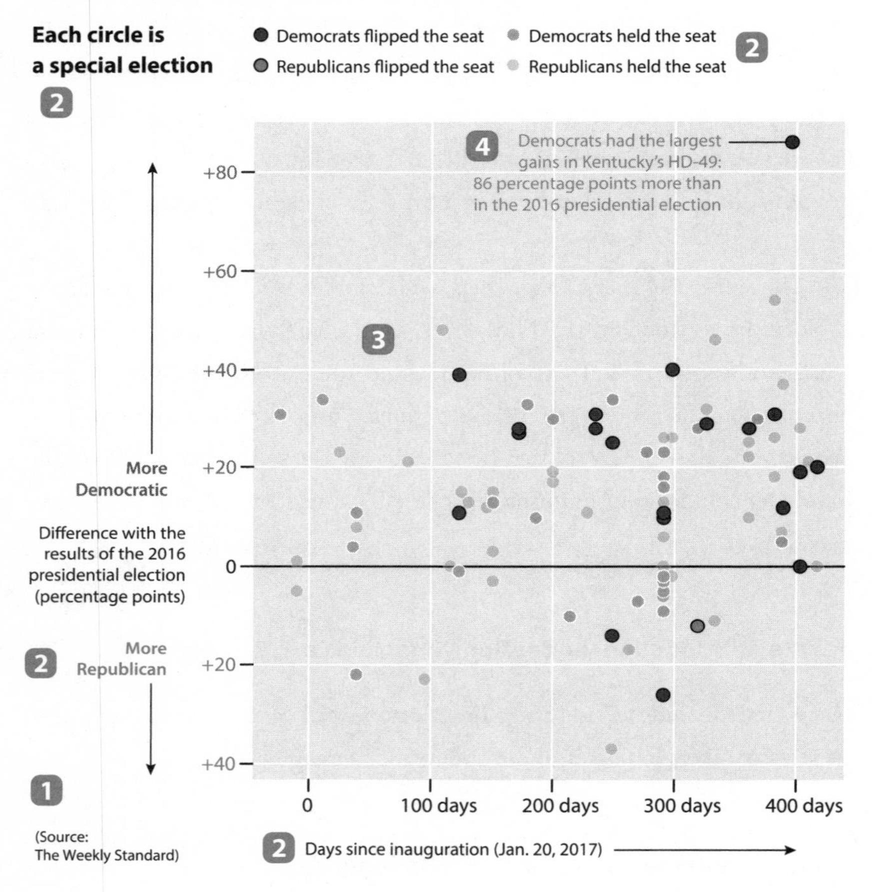

3. Methods of visual encoding

We've already spotted one: color. Grey indicates a Democratic victory; and red, a Republican victory. The color shade, darker or lighter, corresponds to whether either party flipped the seat.

The second method of encoding is position: position on the vertical axis is proportional to the difference in percentage points in comparison to the results of the 2016 presidential election. In other words, the higher above the zero baseline a circle is, the better Democrats did in comparison to 2016; the opposite is true if a circle is below the baseline.

To be even clearer: Imagine that in one of these districts Republicans got 60% of the vote in the 2016 presidential election and that, on this chart, this district's circle sits on the +20 line above the baseline. This means that Republicans got just 40% of the vote in the current special election—that is, a +20 percentage point change in favor of Democrats (if there weren't third-party candidates, of course).

4. Read annotations

Sometimes designers add short textual explanations to charts to emphasize the main takeaways or highlight the most relevant points. In this chart, you can see a note about a race in Kentucky's House District 49, where the Democrats flipped the seat after getting a whopping 86 percentage point increase in support.

5. Take a bird's-eye view to spot patterns, trends, and relationships

Once you have figured out the mechanics of a complex chart like this, it's time to zoom out and think of patterns, trends, or relationships the chart may be revealing. When taking this bird's-eye view, we stop focusing on individual symbols—circles, in this case—and instead try to see them as clusters. Here are a few facts I perceived:

- Since January 20, 2017, Democrats flipped many more seats from Republicans than Republicans from Democrats. In fact, Republicans flipped just one seat.

- In spite of that, both Democrats and Republicans held on to many seats.
- There are many more dots above the zero baseline than below it. This means that Democrats made big gains in the first 400 days after Inauguration Day, in comparison to their results in the 2016 presidential election.

How long did it take me to see all that? Much longer than you may think. Nevertheless, this is not a sign of a poorly constructed chart.

Many of us learned in school that *all* charts must be amenable to being decoded at a quick glance, but this is often unrealistic. Some elementary graphs and maps are indeed quick to read, but many others, particularly those that contain rich and deep messages, may require time and effort, which will pay off if the chart is well designed. Many charts can't be simple because the stories they tell aren't simple. What we readers can ask designers, though, is that they don't make charts more complicated than they should be for no good reason.

In any case, and to continue with the analogy between charts and text I began a few pages back: you can't assume you'll interpret a news story or an essay well if you read its title alone or if you browse it inattentively or hastily. To extract meaning from an essay, you must read it from beginning to end. Charts are the same. If you want to get the most from them, you need to take a deep dive.

Now that we know how to read charts at a symbolic and grammatical level, defending ourselves against faulty ones becomes easier, and we can move to the semantic level of how to interpret them correctly. A chart may lie because:

- It's poorly designed.
- It uses the wrong data.

- It shows an inappropriate amount of data—either too little or too much.
- It conceals or confuses uncertainty.
- It suggests misleading patterns.
- It panders to our expectations or prejudices.

If charts are based on representing data through different methods of encoding as faithfully as possible, then it won't be surprising if I tell you that breaking this core principle will invariably lead to visual lies. We turn to them now.

Chapter 2

Charts That Lie by Being Poorly Designed

Many things can go wrong when designing a chart. Maybe the size of the symbols representing the data isn't proportional to the data itself. Or it may happen that the scales of measurement are off, or even that they were chosen without a sound understanding of the nature of the numbers we're trying to convey.

Now that we've learned the core principles of chartmaking, we're ready to see what happens when those principles are broken.

Partisanship may be inevitable in politics, but it isn't an excuse for shoddy charting. On Tuesday, September 29, 2015, Congress held a hearing with Planned Parenthood's former president Cecile Richards. Planned Parenthood is a U.S. nonprofit organization that provides reproductive health care and sex education. Conservative Republicans have often attacked it because it offers abortions among those services.

Utah Republican Jason Chaffetz showed a chart like this during a heated exchange with Richards.[1] Don't bother trying to read the figures yet; they are written in tiny type, as on the original:

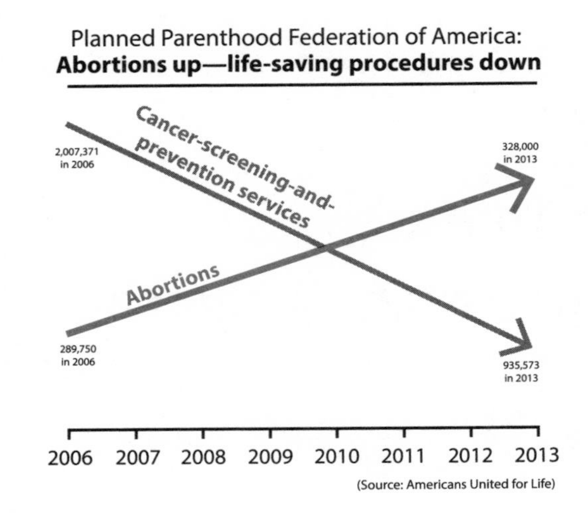

Chaffetz challenged Richards to look at the chart, and he demanded a response from her. Richards, who wasn't sitting close to the screen where the chart was being projected, squinted and looked puzzled. Chaffetz said then, "In [gray], that's the reduction in the breast exams, and the red is the increase in the abortions. That's what's going on in your organization."

Richards replied that she didn't know where the chart had come from and added that, in any case, "it does not reflect what happens at Planned Parenthood."

Chaffetz exploded: "You're going to deny that we've taken those numbers out of your report? . . . I pulled those numbers directly out of your corporate reports!"

That's only partially true, and Richards pointed it out: "The source for this [chart] is actually Americans United for Life (AUL), which is an antiabortion group, so I would check your source." Chaffetz stuttered a bit: "We . . . We will get to the bottom of the truth of that."

The "bottom of the truth" is that the numbers on this chart were indeed pulled out of Planned Parenthood's report, but AUL's representation of those numbers is twisted. The chart emphasizes that the number of cancer-screening-and-prevention services declined at the same rate as the increase in abortions. This is false. The chart lies because it's using a different vertical scale for each variable. It makes it appear that in the most recent year, 2013, Planned Parenthood conducted more abortions than prevention procedures.

Try now to read the minuscule numbers. There was a sharp drop in cancer-screening-and-prevention services, from two to one million, but the increase of abortions was just from roughly 290,000 to 328,000. If we plot the figures using a common scale, here's what we get:

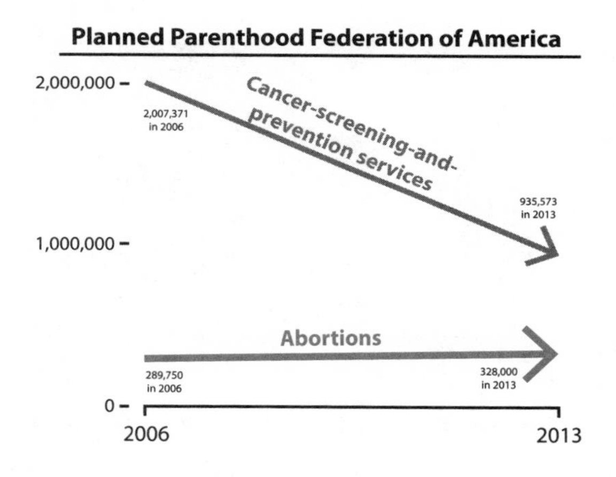

PolitiFact, an excellent fact-checking website, did some inquiries into how the original chart came about and interviewed several sources that could help explain variations in the types of services Planned Parenthood provides:[2]

> The number of services provided in each category tend[s] to fluctuate from year to year for myriad reasons, ranging from changes in laws and medical practices to the opening or closure of Planned Parenthood clinics.

And it's not just that the increase in number of abortions is almost imperceptible; it's that it has declined slightly since 2011. How so, if the numbers on the original chart are correct? The reason is that, even if all years between 2006 and 2013 are labeled on the original chart, the lines compare only the figures for 2006 and those for 2013, ignoring what happened in other years. Here's a year-by-year chart for number of abortion procedures, revealing two tiny peaks in 2009 and 2011:

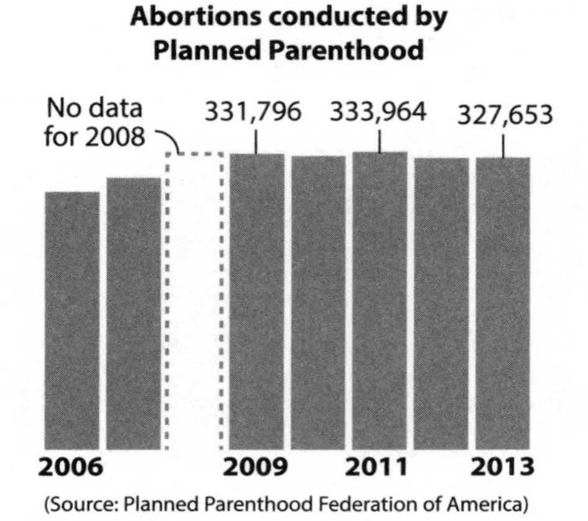

Therefore, Americans United for Life was not only distorting the representation of the data—the theme of this chapter—but also concealing important information, a problem we'll return to in chapter 4.

Data scientist and designer Emily Schuch collected Planned Parenthood's annual reports between 2006 and 2013—with the exception of 2008—and showed that the organization does many more things other than prevention services and abortions. It also does pregnancy and prenatal care, testing for sexually transmitted diseases, and other services. Abortions are just a tiny percentage of everything Planned Parenthood does. You can see Schuch's chart on the facing page.

Schuch revealed that STI/STD services (testing and treating sexually transmitted infections and diseases) increased by 50% between 2006 and 2013. She also investigated why cancer screenings decreased so much in the same period, and she found this tentative explanation:

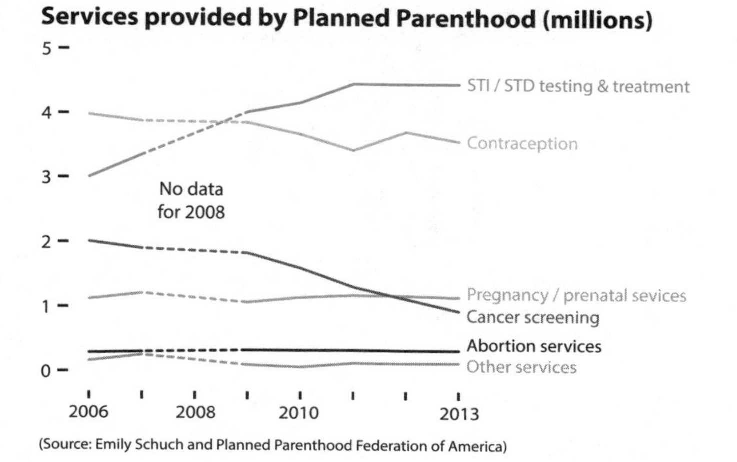

> National guidelines for the frequency of cervical cancer screenings were officially changed in 2012, but the American College of Obstetricians and Gynecologists started advising a reduction of frequency in 2009. **Previously, women received a pap smear to screen for cervical cancer once a year, but it is now recommended that pap smears should be performed every three years.**[3]

For our purposes in this book, it is irrelevant whether you support or oppose public funding of Planned Parenthood. Schuch's chart is objectively better than AUL's because it includes all relevant data and it doesn't skew the representation of that data to push an agenda. That's the difference between designers who draw charts to inform civil and honest discussions and those who design sloppy propaganda.

Visual distortions are a constant source of good laughs for those who can read and design charts appropriately, but they can also be infuriating. Suppose that I want to inform the public of the success of my company in comparison to my main competitors, and I say it with a chart:

My company also dominates the market. Just take a look at our enormous share of all sales!

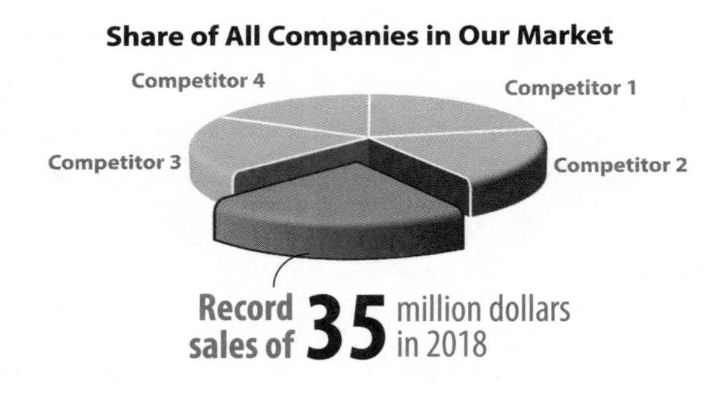

And our business has been booming since 2011!

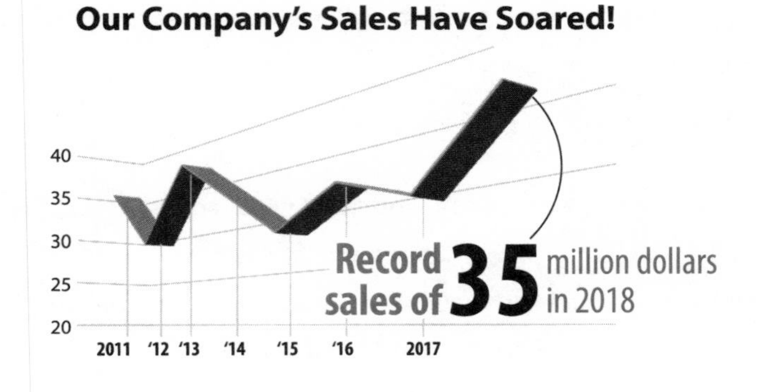

Three-dimensional perspective effects are the scourges of the chart world. You may think I'm exaggerating with these fictional examples. I'm not. Browse through the press releases, presentation slides, websites, or reports of many organizations and you'll find graphics like these—or worse. They look flashy and dramatic, but they fail miserably at informing you.

Try to see whether my claims of market dominance and increasing sales have merit. It's hard, isn't it? By choosing convenient angles, I have exaggerated my success (it would be different, by the way, if these charts were interactive or seen through a virtual reality device, as in those cases the reader could move around the 3-D charts to see them in two dimensions).

Some people contend that 3-D effects are fine because you can always write every single figure that your chart represents on top of your bars, lines, or pie segments; but then, what is the point of designing the chart in the first place? A good graphic should let you visualize trends and patterns without having to read all the numbers.

If I remove the exaggerated perspective, the height of the bars will be proportional to the data, as will the area of the pie chart segments and the height of the line. You would see quite clearly that Competitor 1 is slightly more successful than I am and that our sales in 2018 were lower than our peak in 2013.

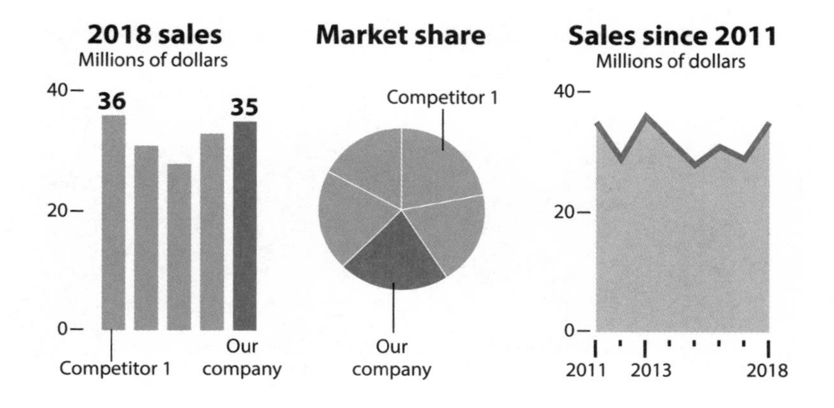

Chart distortions, then, are usually the result of fudging with scales and proportions. In December of 2015, the Obama White House tweeted, "Good news: America's high school graduation rate has increased to an all-time high," with this chart:[4]

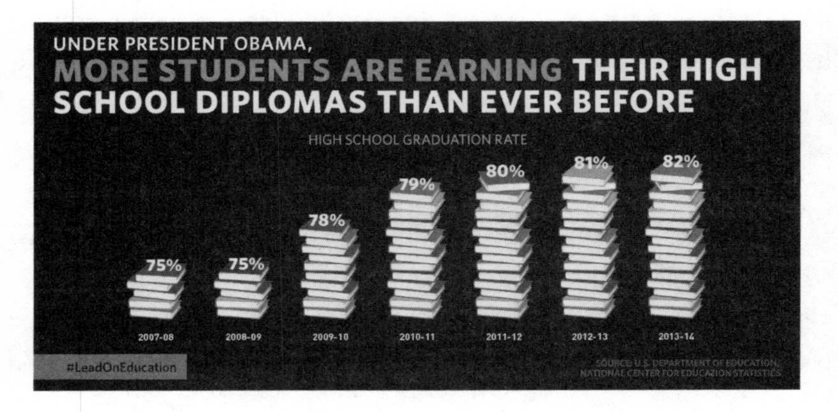

The design of a graphic—its scales and encodings—should depend on the nature of the data. In this case, it's a yearly percentage, and the method of encoding is height. For that reason, it's best to make the height of the bars proportional to the numbers by putting the baseline at 0% and the upper edge at 100%:

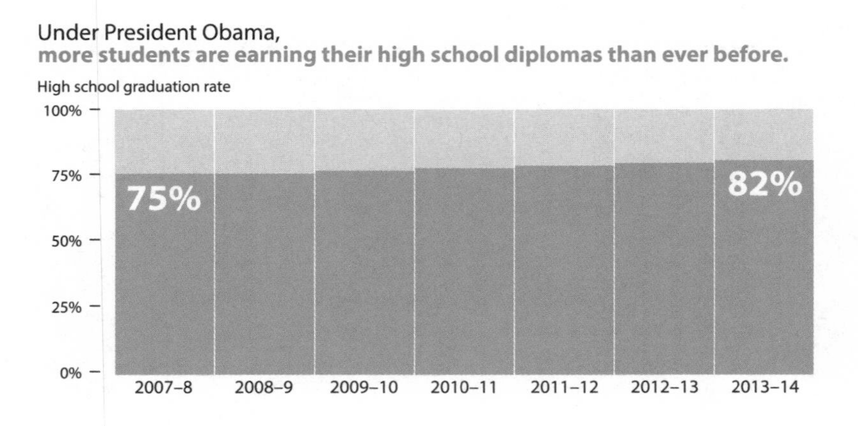

This chart makes bar height proportional to the data, and by using a big type size for the initial and final rates, it preserves an important feature from the original one: emphasizing that a 7 percentage point increase in graduation rates is great news.

The White House chart is problematic because it truncates both the vertical axis (*y*), and the horizontal axis (*x*). As was pointed out by news website Quartz (https://qz.com) on the basis of U.S. Department of Education data, by starting the *x* axis with the 2007–2008 school year, the author hid the fact that high school graduation rates have been increasing since the mid-1990s, not just under Obama:[5]

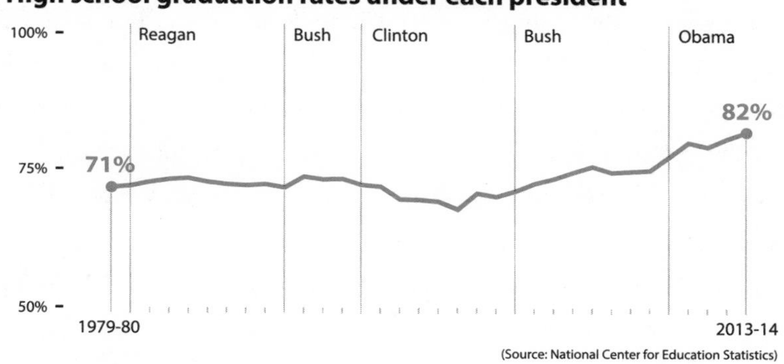

You may be wondering why I didn't put the baseline at zero on this chart. I'll have much more to say about picking baselines later in this chapter, but the main reason is that I usually advise a baseline of zero when the method of encoding is height or length. If the encoding is different, a zero baseline may not always be necessary.

The encodings in a line chart are position and angle, and these don't get distorted if we set a baseline that is closer to the first data point. The lines in the following two charts look exactly the same, and neither lies. The only difference is the emphasis on the bottom baseline. On the first chart I emphasized the baseline because it's zero. On the second chart, the baseline looks like any other gridline because I want to make clear that *it's not zero*:

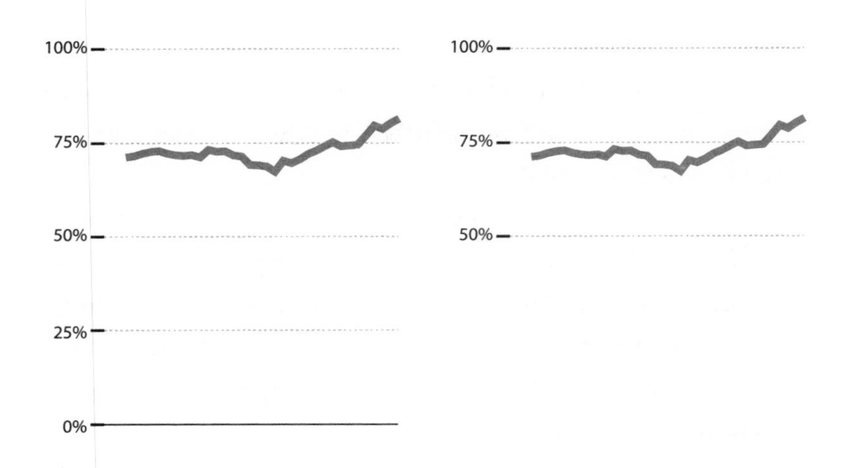

Focusing on the scaffolding of a chart—its scales and legends—before decoding its content will help identify distortions. Below is a chart published in 2014 by the Spanish city of Alcorcón to celebrate a spectacular job market under the current mayor, David Pérez García. The two sides are like mirror images: unemployment went up sharply under the previous mayor, Enrique Cascallana Gallastegui, and it went down at the exact same rate with the current mayor, Mayor Pérez García—or so it seems, until you read the tiny labels.

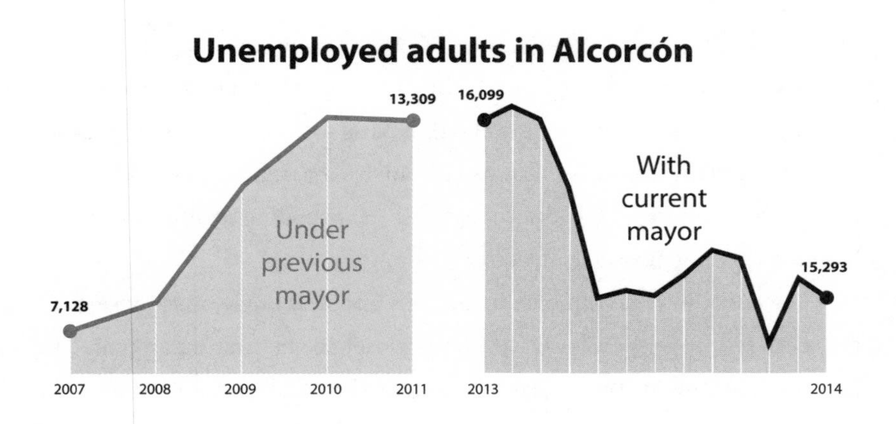

The challenge here is that both vertical and horizontal scales are wrong. The first half of the graphic displays yearly data while the second shows monthly data. When we put both halves of the line on the same vertical and horizontal scales, the drop in unemployment still looks like a positive story, but certainly a less spectacular one than before.

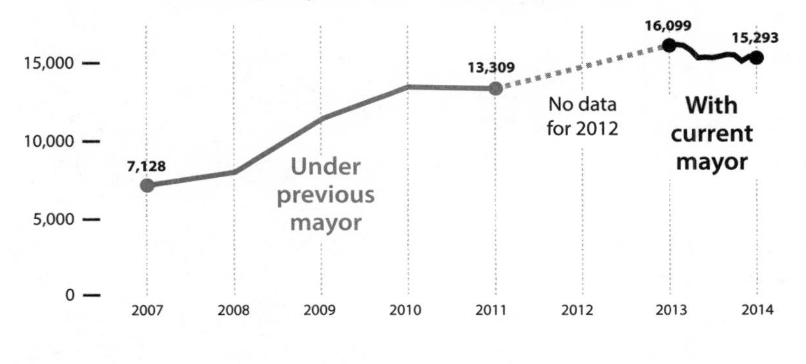

You may think that it's harmless to torture the proportions of charts and use inconsistent scales—either inadvertently or on purpose. After all, as I've heard some chart designers say, "everybody should read labels and scales. If they do so, they can mentally undistort the chart." Sure, I agree that we need to pay attention to labels. But why mangle proportions and scales and make readers' lives miserable?

Moreover, even if we *do* pay attention to the scaffolding of a misshapen chart and make an effort to visualize the proportions correctly inside our heads, the chart may unconsciously bias our perceptions anyway.

A group of researchers from New York University designed two versions of several charts about a fictional variable: access to drinking water in the imaginary towns of Willowtown and Silvatown.[6] The first version of each chart depicted data correctly and did not distort scales and proportions. The second version did distort scales and proportions: it used a bar chart with a truncated vertical axis, a bubble chart in which the numbers aren't proportional to the circle area, and a line chart with an aspect ratio that minimized the change depicted. Here are three pairs of correct versus deceptive charts:

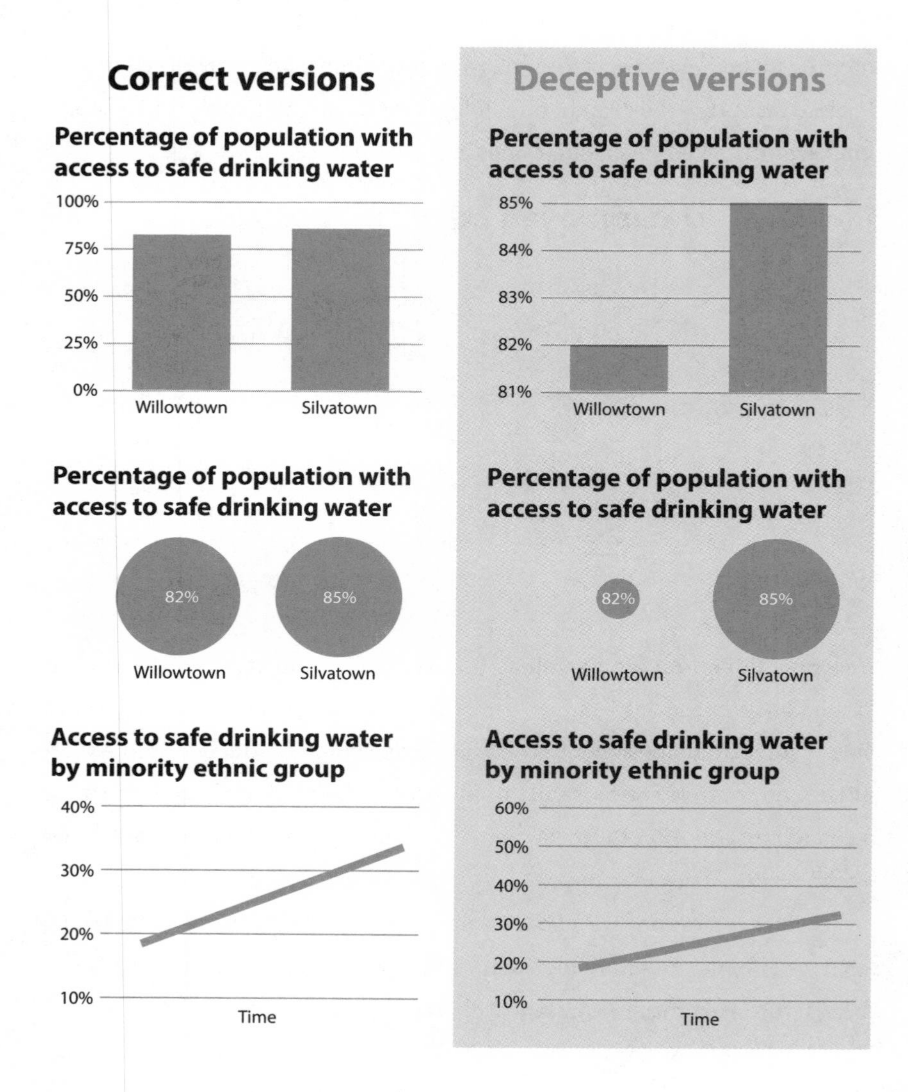

Researchers asked several groups of people to compare the objects on the charts—"Is the second quantity slightly bigger or much bigger than the first one?"—and the results showed that even if people could read the labels on the scales or the numbers themselves, they were still misled. Those readers who were more educated and used to seeing these kinds of charts did slightly better, but they got them wrong anyway.

Before scholars began conducting experiments like this, some bad actors already had an intuitive understanding of chart deception techniques. In December 2015, *National Review* magazine quoted the blog Power Line with the headline "The Only Climate Change Chart You Need to See."[7] Unfortunately for *National Review*, it seems to me that Power Line pranked them with that chart:

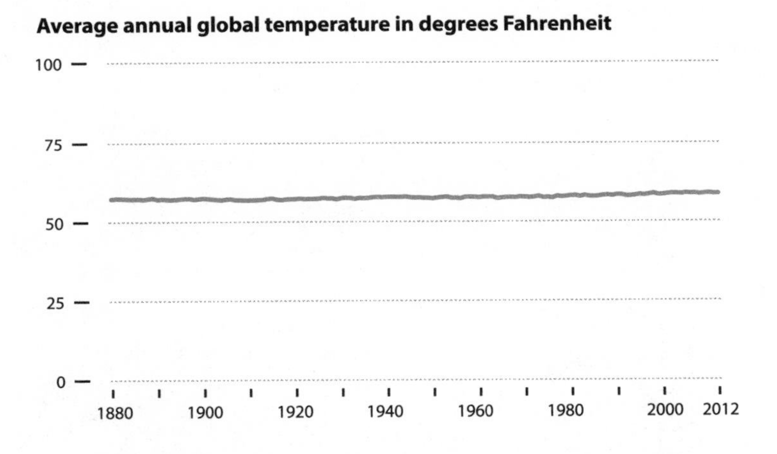

Data analyst Sean McElwee and many other people made fun of the chart on social media. McElwee wrote on Twitter: "No need to worry about the national debt then either!" and inserted a chart like this:[8]

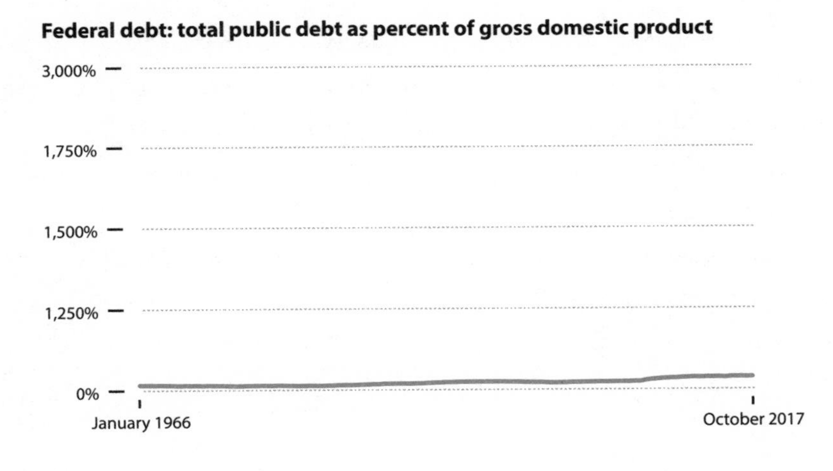

I used to be worried that the U.S. debt reached 103% of gross domestic product in October 2017, but this chart suggests that my concern may have been exaggerated. We're still so far from 3,000%!

Richard Reiss, a research fellow at the City University of New York's Institute for Sustainable Cities, added some tongue-in-cheek annotations to the original chart that reveal one of the many reasons why the choice of scales is so wrong:

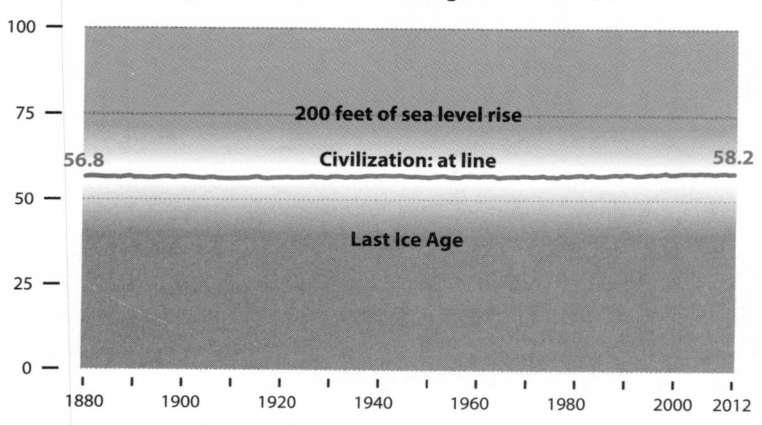

Reiss's joke contains a lot of wisdom. The difference between the beginning and the end of the line is 1.4°F, or nearly 0.8°C. That's a very significant variation, even if it sounds small in absolute terms. During the Little Ice Age, between the 15th and the 19th centuries in the Northern Hemisphere, the average global temperature was roughly 1°F lower than late 20th-century levels,[9] but the consequences were dramatic, as colder weather contributed to famines and epidemics.

Consequences could be equally grim, or worse, if global temperatures were to increase by 2°F or 3°F in the next 50 years—an entirely realistic estimate. If average temperatures reached 100°F—the absurd upper boundary of Power Line's chart—Earth would become a pit of hell.

Moreover, Power Line's designer set the baseline at zero. This is ludicrous for several reasons, the main one being that the Fahrenheit and Celsius temperature scales don't have a minimum of zero (only the Kelvin scale does).

A chart designer intending to inform us, not to mislead us, should take all this into account and choose sensible scales and baselines:

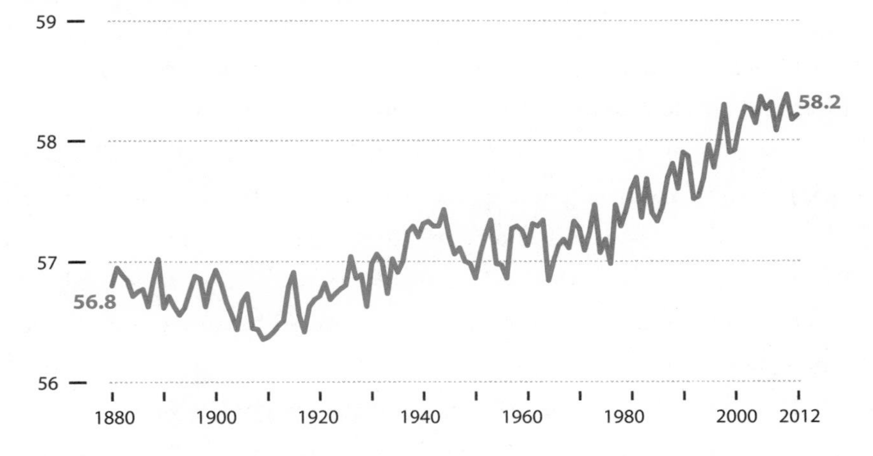

You may have heard some people say that "all charts should start at zero!" This notion was popularized by books such as Darrell Huff's 1954 *How to Lie with Statistics*. I hope that this example has disabused you of that notion. Huff's book, as old as it is, contains much good advice, but this is an exception.

Chart design, like writing, is as much a science as it is an art. There aren't many rules that are set in stone; instead, what we have are mostly flexible principles and guidelines with numerous exceptions and caveats. As chart readers, should we demand that all charts begin at zero? Well, that depends on the nature of the information, on the space available for the chart, and also on the choice of encoding.

Sometimes these considerations clash with each other. Here's a chart of the average life expectancy at birth in the world. It doesn't look like a big change, does it?

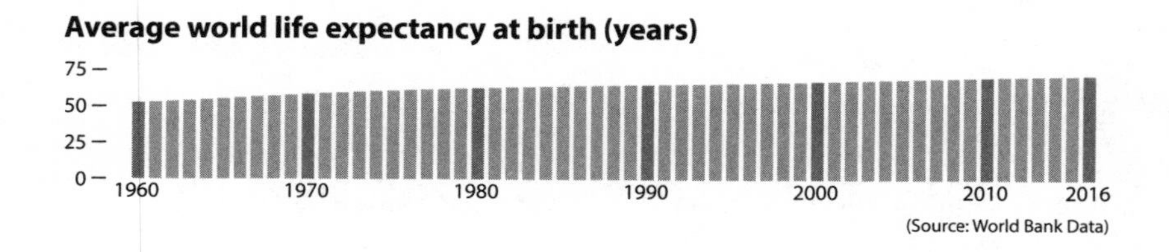

I faced two challenges with this chart: I had a very wide and short space for it, and I decided to use height (a bar chart) to encode my data.

The combination of those two factors led to a chart that flattens a change that in reality is quite impressive: the average life expectancy in the world was 53 years in 1960, and 72 in 2016. That's a 35% increase. But the chart doesn't emphasize this, as bars should start at zero, so their height is proportional to the data, and they are short overall because of the aspect ratio of the chart.

There are never perfect solutions in chart design, but we can find a reasonable compromise if we reason about the data itself. This data set of life expectancy of all countries in the world has a possible zero baseline. It's possible, of course, but not *logical*. If any country ever had a life expectancy of zero, it would mean that all children being born in that country would die as soon as they got out of the womb.

Therefore, setting the baseline at zero—something advisable in a bar chart—leads to a dubious graphic in this particular case. This is the contradiction I mentioned before: the method of encoding (height) forces us to do one thing, but the data itself suggests that we should do something different.

The compromise I'd choose would be to not use height to encode the data; instead, I'd try position and angle, as on a line chart. Then, I'd set the baseline at a value closer to the initial one, like this:

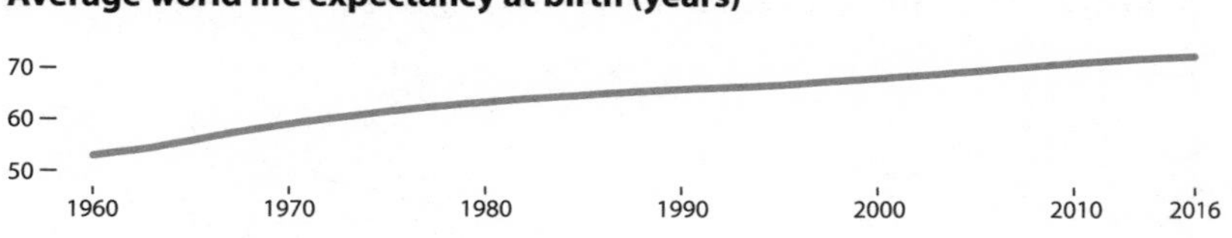

The aspect ratio—the proportion between the chart's width and height—is not ideal, but we have the world we have, not the one we'd like to have, and I had a very short and wide space for my chart. Journalists and chart designers always deal with trade-offs, and all that we can ask from them as readers is that they be honest when weighing their options.

However, if designers aren't constrained by the space available, we can demand that they not do any of these:

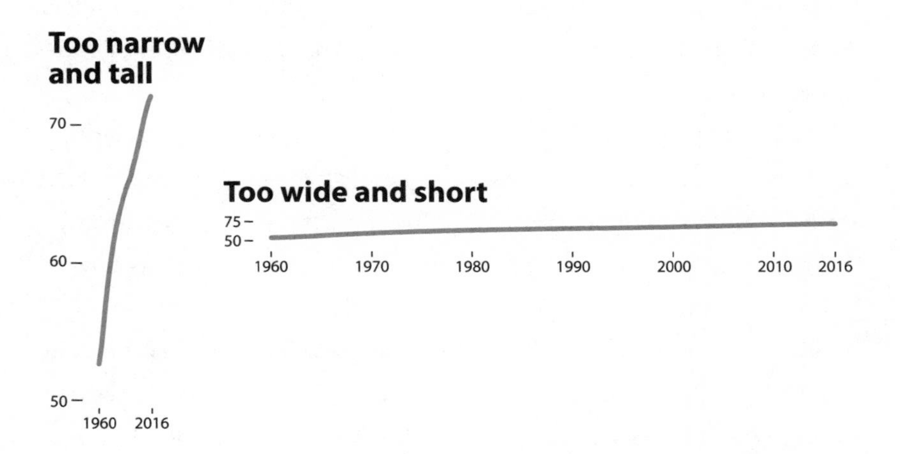

Instead, they should design a chart that finds an aspect ratio that neither exaggerates nor minimizes the change. How is that so? We are representing a 35% increase. That's 35 over 100, or 1/3 (aspect ratios put width first, so in this case it would be 3:1). I can approximate the proportions of the chart to this: three times as wide as the line is tall. You can see the result on the next page.

There's an important caveat: this isn't a universal rule of chart design. Recall what I mentioned before about considering not numbers in the

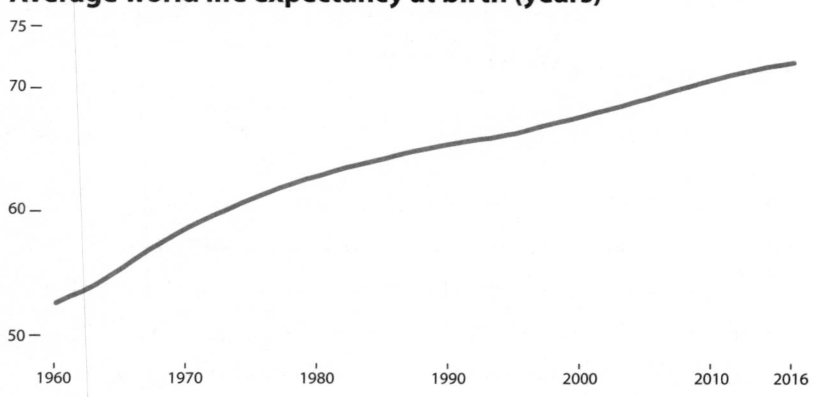

abstract but rather what they mean. Sometimes an increase of 2%, as in the case of global temperatures, can be very significant; however, it'll look deceptively small if the chart has an aspect ratio of 100:2, or 50:1—50 times as wide as it is tall.

One core theme in this book is that chart design is similar to writing. Decoding a chart is similar to reading a text, although chart reading isn't always linear, as traditional reading is. To extend the analogy, we can argue that the "too narrow and tall" chart we saw on the previous page is a hyperbole and that the "too wide and short" one is an understatement.

As in writing, in charting it can also be debatable whether a certain claim is a hyperbole, an understatement, or a reasonable middle point between the two. Similarly, chart design can't be based on absolute rules, but these rules aren't arbitrary either. We can reach an adequate—but never perfect—compromise if we apply the elementary grammatical principles we learned in chapter 1 and then reason about the nature of the data we're handling.

Some charts may look distorted but truly are not. For example, look at the chart below, but don't read the scale labels. Focus just on the circles, each corresponding to a country. The chart shows life expectancy country by country (position on the vertical axis) and gross domestic product per capita (position on the horizontal axis):

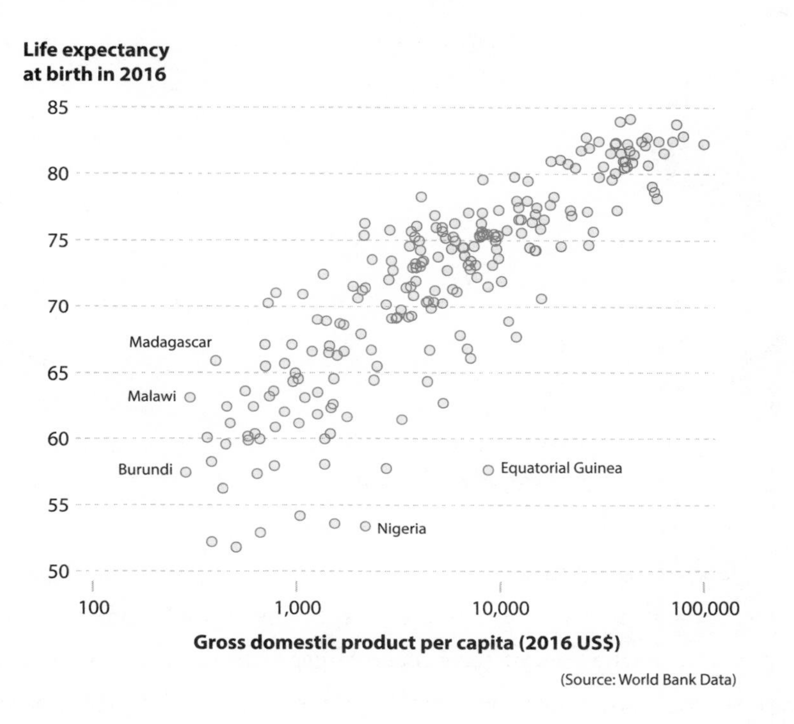

Now read the axis labels. Do you notice that the horizontal scale looks odd? Instead of even intervals (1,000; 2,000; 3,000), the labels are powers of 10: 100; 1,000; 10,000; and 100,000. This kind of scale is called a logarithmic scale—or, more precisely, a base-10 logarithmic scale (other bases are possible).

"This chart is a lie!" you may have yelled, clenched fist shaking in the air. Not so fast, I'd argue. Let's think about the data and what this chart intends to show. (Hint: the reason I chose that scale is related to the countries I highlighted.)

Let's see the data on a horizontal scale with even intervals. This is called an arithmetic scale, the most common in charts of all kinds:

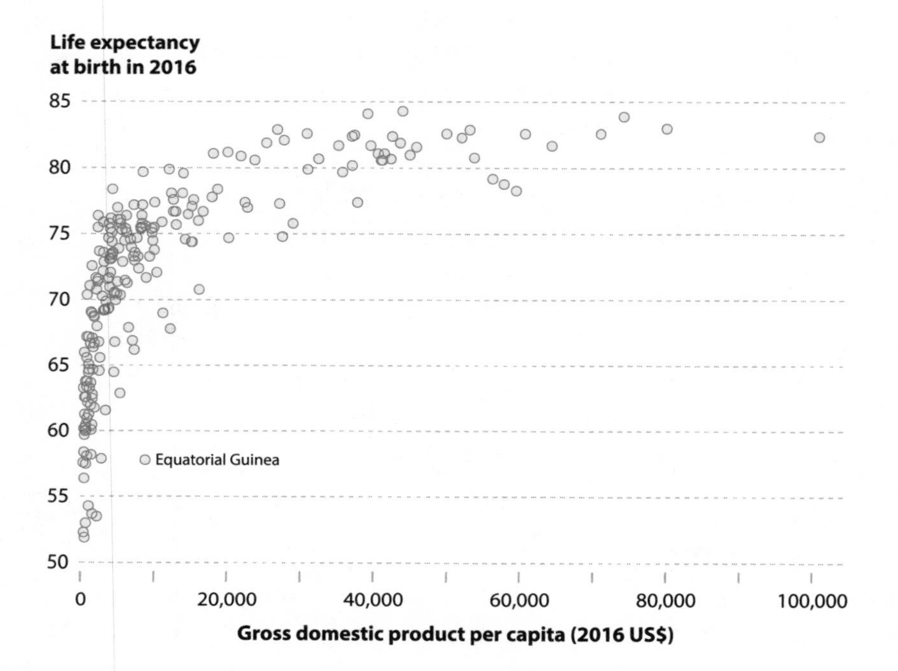

I labeled several African countries on the first version of the chart. Try to locate them on this second one. You may be able to see Equatorial Guinea because it's such a special case: much larger GDP per capita than other nations with similar life expectancies. But the countries I wanted to scrutinize and that I highlighted on my first chart—Nigeria, Malawi, Madagascar, Burundi—have such low GDPs per capita that they get enmeshed with all other poor nations with relatively low life expectancies.

In the same way that we shouldn't automatically believe a chart before reading it carefully, neither should we rush to call a chart a lie before we think about what it was designed for. Recall the example that opens this book, the county-level map of presidential election results. That graphic is correctly built, and it wouldn't lie if our goal were to reveal geographic voting patterns. However, it was used to represent the number of people who voted for each candidate, and it fails miserably at that.

Neither of the scatter plots I've just shown you can be called a lie without assessing its purpose. Is the goal to show the relationship between GDP per capita and life expectancy? Then the second one may be better. It lets us see that in a pattern that looks like an inverted L—many countries with low GDP per capita and high variability of life expectancies (the dots on the vertical arm of the inverted L), and another group of richer countries with high variability of GDP per capita and low variability of life expectancies (the horizontal arm of the L):

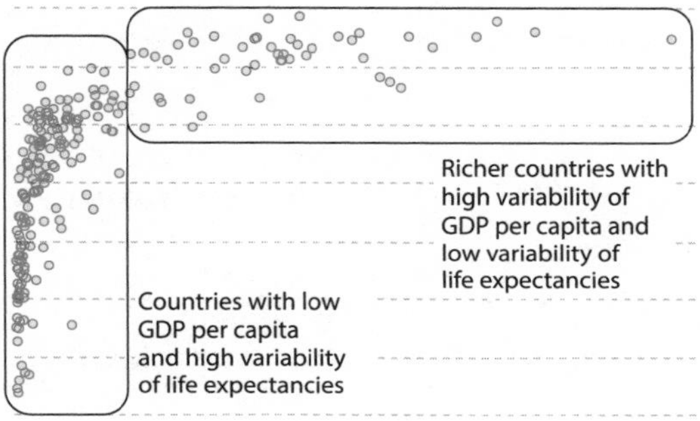

But this wasn't the purpose of my first chart. What I wished to emphasize is that there are certain African countries that have low life expectancies even if their GDP per capita is relatively high—Equatorial Guinea and Nigeria—and others that, in comparison, are very poor and have relatively high life expectancy, such as Malawi, Burundi, and especially Madagascar. If I do this with an arithmetic scale, most of these countries are impossible to spot.

Logarithmic scales sound complicated, but you're likely familiar with some examples. The Richter scale, which measures the strength of earthquakes, is a base-10 logarithmic scale. This means that an earthquake with a magnitude of two on the Richter scale is not twice as strong as one with a magnitude of one. It's 10 times stronger.

Logarithmic scales are also used to plot exponential growth. Imagine that I have four gerbils in my backyard, two males and two females, and that they mate. Each gerbil couple gives birth to four little gerbils that go on to mate with the little children of the other couple. Each subsequent couple of cute rodents gives birth to four little ones, systematically.

I could plot the growth of the gerbil population this way:

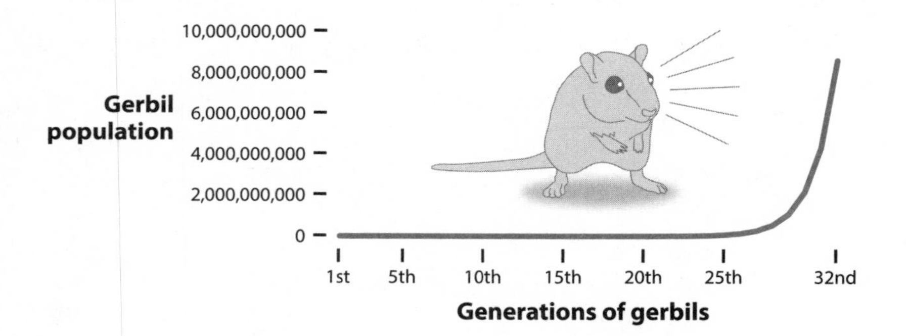

If I decided how much gerbil food to buy based on this chart, I'd think that I wouldn't need to change my investment much until the 25th generation. The line remains flat until then.

But what the chart conceals is that the number of gerbils is doubling in every generation, so I'll need to double the food. A logarithmic scale with base two—where each increment is twice the size of the previous one—may be more appropriate, as I'm interested in the rate of change, not in the raw or absolute change itself. By the 32nd generation, there will be more gerbils living in my backyard than people in the world, so I may also want to do something about birth control:

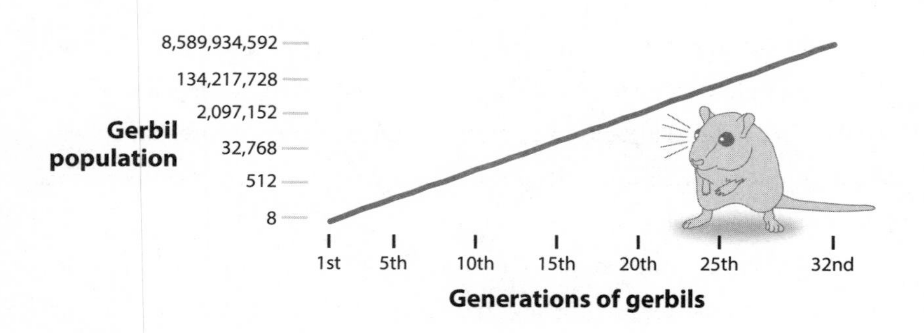

Many charts lie not because their scales are arithmetic or logarithmic but because the objects that encode the data are themselves truncated or twisted in odd manners. I've seen plenty of charts that cut off axes and symbols like this:

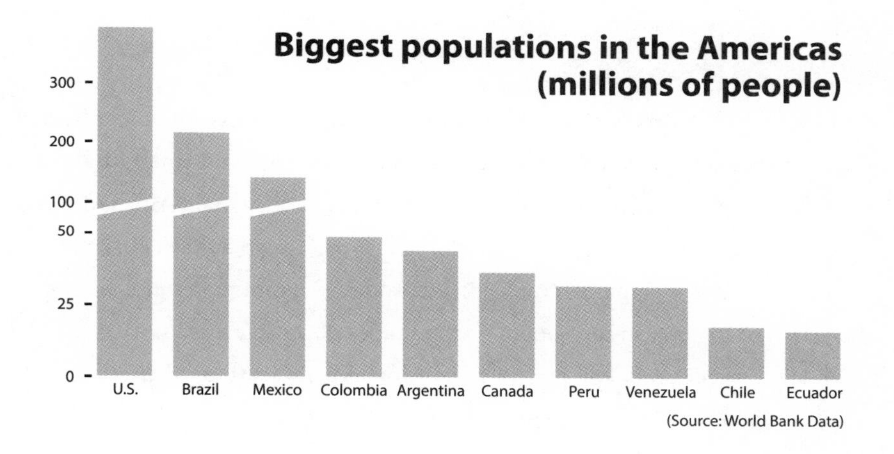

This chart lies because the intervals on the vertical scale aren't even and the three first bars are truncated. The real proportions look like this:

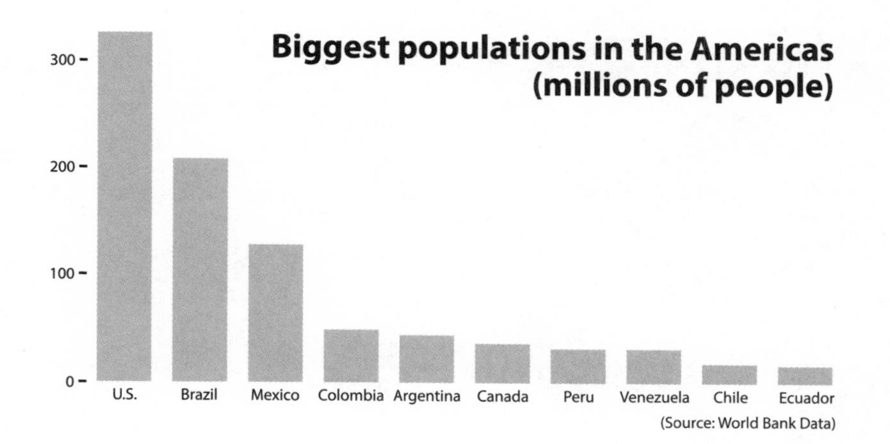

The correct version of the chart has its shortcomings, though. For instance, we may complain that it's now much harder to tell whether there are any differences among the smaller countries. As readers, we could demand from the designer that instead of being shown just one chart we be shown two: one with all countries on the same scale, and another one zooming in to the less populous ones. That way, all goals would be achieved and scale consistency would be preserved.

All maps lie, claims cartographer Mark Monmonier in his canonical book *How to Lie with Maps*. This is a mantra that we may extend to all kinds of charts—although, of course, not all lies are created equal. All maps lie because they are based on the principle of projecting a spherical surface, the Earth, onto a plane. All maps distort some geographic feature, such as the sizes of the areas represented or the shapes of those areas.

The projection below is called the Mercator projection, after its 16th-century creator. It makes regions far from the equator much bigger than they truly are—for instance, Greenland isn't larger than South America, and while Alaska is enormous, it's not *that* enormous—but it preserves the shapes of landmasses.

Another projection, called Lambert's Cylindrical Equal Area, sacrifices shape accuracy to make landmass area proportional to reality:

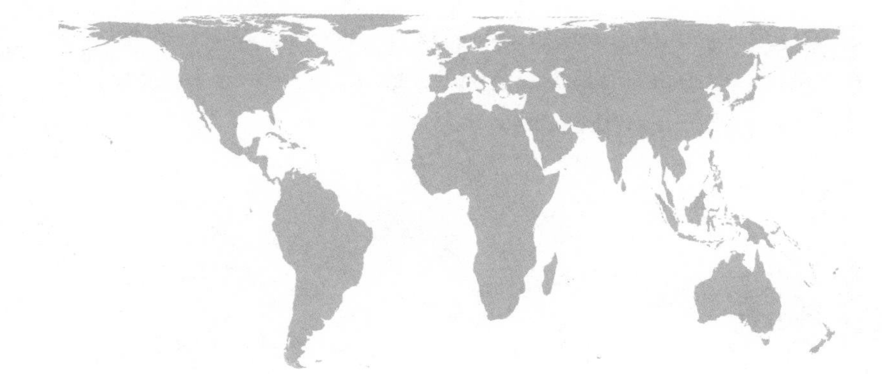

The following projection, called Robinson's, preserves neither shape nor area. Instead, it sacrifices a bit of both to achieve a balance that's more pleasing to the eye than Lambert's:

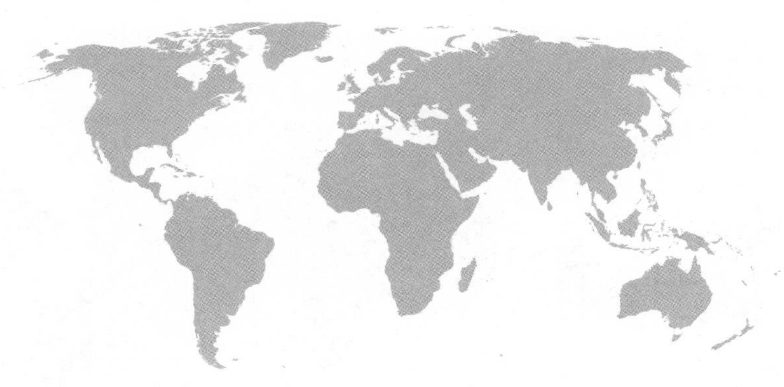

As with charts that are honestly designed, no map projection is good or bad in the abstract. It is *better* or *worse*, depending on the goal of the map. If you want to hang a world map in your child's room, Robinson's projection is more educational than either Mercator's or Lambert's. But if your goal is to have a tool for sea navigation, the Mercator projection is more suitable—that is what it was designed for, after all.[10]

Although all map projections are lies, we know that they are white lies, consequences of the fact that any chart is a limited and imperfect representation of reality, not reality itself. All charts suffer from that limitation.

Maps may also lie because of bad design choices, intended or otherwise. For instance, I could mess around with color scales to prove either that poverty is a very localized problem in the United States . . .

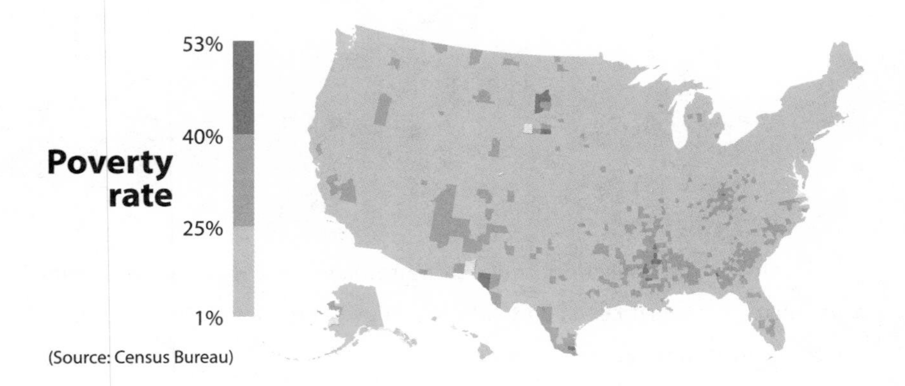

. . . or that poverty is an enormous challenge everywhere:

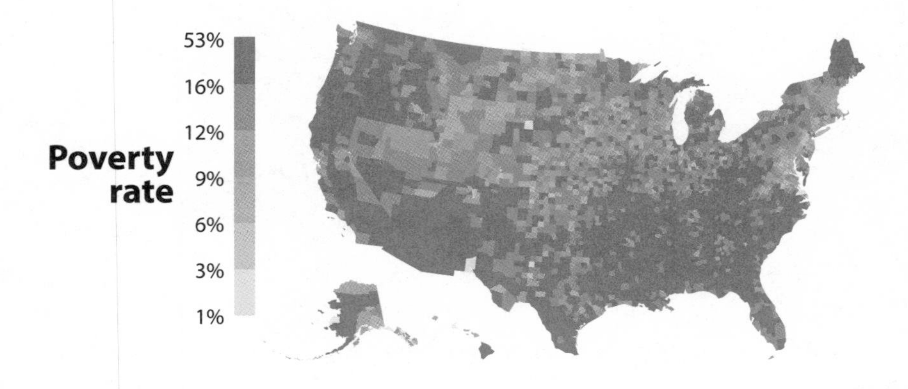

These two maps bias your perception because I carefully chose the color groups (or bins) to produce an understatement (first map) or a hyperbole (second map). The problem with the second map is the color scale: the darkest shade is used for counties that have a poverty rate of between 16% and

53%. Half the counties in the United States have those levels of poverty; the other half have poverty rates of between 1% and 16%. That's why the map looks alarmingly red.

A more sensible scale would distribute the counties evenly through the different color bins. There are around 3,000 counties in the United States. On the map below, each of the six color bins contains around 500 counties (3,000 counties divided by six bins equals 500 counties by color bin):

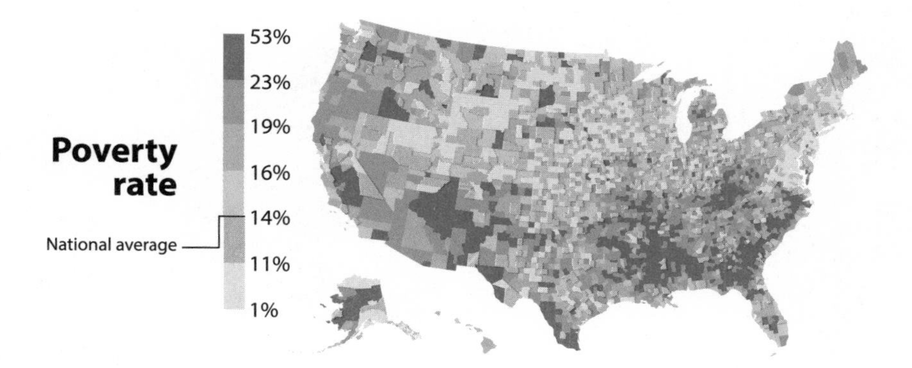

But wait! Imagine that the title of my map were "Counties with Poverty Rates Higher Than 25%." In that case, my first map would be appropriate, as it emphasizes just counties with poverty rates higher than 25% and 40%. As we've seen, chart design depends on the nature of the data being shown and on the insights that we want to extract from it.

The quality of a chart depends on its encoding the data accurately and proportionally, but a prior consideration is the reliability of the data itself. One of the first features to look at in a chart is its source. Where did the data come from? Does the source look trustworthy? How can we assess the quality of the information being shown to us? We move to that topic in the next chapter.

Chapter 3

Charts That Lie by Displaying Dubious Data

One of my favorite mantras is "Garbage in, garbage out." It's a popular saying among computer scientists, logicians, and statisticians. In essence, an argument may sound very solid and convincing, but if its premise is wrong, then the argument is wrong.

Charts are the same way. A chart may look pretty, intriguing, or surprising, but if it encodes faulty data, then it's a chart that lies.

Let's see how to spot the garbage before it gets in.

If you like charts, social media can be an endless source of exciting surprises. A while ago, mathematician and cartographer Jakub Marian published a map displaying the density of heavy metal bands in Europe. Below is my own version of it, highlighting Spain, where I was born, and Finland.[1]

Being a fan of many hard rock and (nonextreme) metal bands, I loved the map and spread the word about it through my contacts on Twitter and Facebook. It confirmed something I've always suspected: many bands are based in northern countries, with Finland being what we could call the Metal Capital of the World.

But then I thought twice and asked myself: Is the source for this map

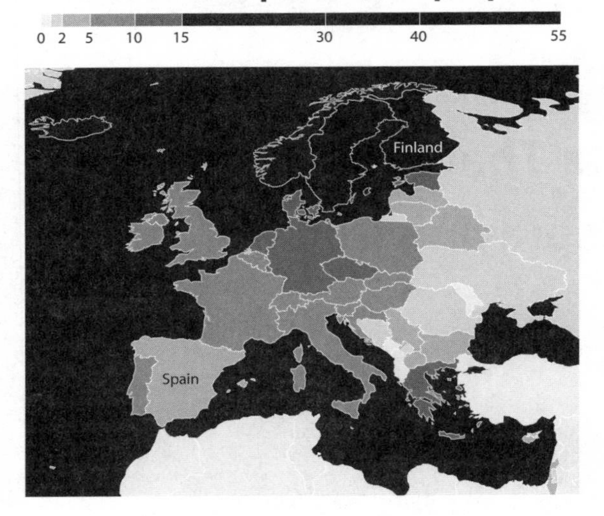

trustworthy? And what does the source mean by "metal"? I felt some skepticism was warranted, as one key lesson of this book is that the charts most likely to mislead us are those that pander to our most deeply held beliefs.

The first thing to look for when reading any chart is whether its author or authors identify sources. If they don't, that's a red flag. We can derive a general rule of media literacy from this:

Distrust *any* publication that doesn't clearly mention or link to the sources of the stories they publish.

Fortunately, Jakub Marian is aware of good transparency practices, and he wrote that the source for his data was a website called the Encyclopaedia Metallum. I visited it to see whether its data set comprised *just* heavy metal bands.

In other words, when checking a source, we must assess *what is being counted*. Is the source counting just "metal" bands or is it counting something else? To do this verification, let's begin by thinking about the most paradigmatic metal band you can evoke, the one that embodies all the values, aesthetics, and style we commonly associate with metal. If all bands in the Encyclopaedia Metallum are more similar than not to this ideal band—if they share more features than not with it—then the source is probably counting just metal bands.

Go ahead, think of a band.

I bet you've thought of Metallica, Black Sabbath, Motörhead, Iron Maiden, or Slayer. Those bands are very metal indeed. Being from Europe and having grown up during the '80s, however, I thought of Judas Priest. These fellows:

Photograph by Jo Hale © Getty Images

Judas Priest has everything that is metal about metal. I think of them as the most metal of the metal bands because they have every single feature metal is recognizable for. To begin with, the clothing, attitude, and visual style: long hair (with the exception of front man Rob Halford, who's bald), tight leather clothes, shiny spikes on black pants and jackets, scowling facial expressions, and defiant poses.

What about the performing and musical features? They are also pure metal. Search for a few Judas Priest video clips, maybe songs such as "Firepower," "Ram It Down," "Painkiller," or "Hell Bent for Leather." You'll notice the endless guitar riffs and solos, the thundering drums, the headbanging—

synchronized, which is even *more* metal—and Halford's vocals, which make him sound like a banshee.

If all the bands in the Encyclopaedia Metallum are more similar than dissimilar to Judas Priest, then this source is counting only metal bands. However, being familiar with the scholarly literature (yes, there's such a thing) and the history of metal as well as the Wikipedia entries about the genre, I've sometimes seen other kinds of bands being called "metal." For instance, these folks, who almost certainly are *not* metal:

Photograph by Denis O'Regan © Getty Images

That's a glam rock group called Poison, which was very popular when I was a teenager. Some sources, including Wikipedia, label it as "metal," but that's quite a stretch, isn't it? I've also seen melodic rock bands such as Journey or Foreigner being called heavy metal in some magazines. Both Journey and Foreigner are fine bands—but *metal*? I don't think so.

In any case, I spent a few minutes taking a look at the Encyclopaedia Metallum's database, and none of those bands were in it. I took a look at some of the tens of thousands of groups listed, and they all look pretty metal to me, at least at a quick glance. I didn't verify the source thoroughly, but at least I made sure that it looks legitimate and it didn't make gross mistakes.

I felt safe to send the map to friends and colleagues.

When reading charts, it is paramount to verify *what* is being counted and *how* it's being counted. My graduate student Luís Melgar, now a reporter based in Washington, DC, did an investigation titled "A School without a Roof" about homeless children enrolled in school in Florida. Their number increased from 29,545 to 71,446 between 2005 and 2014. In some Florida counties, a bit more than one in five students are homeless:

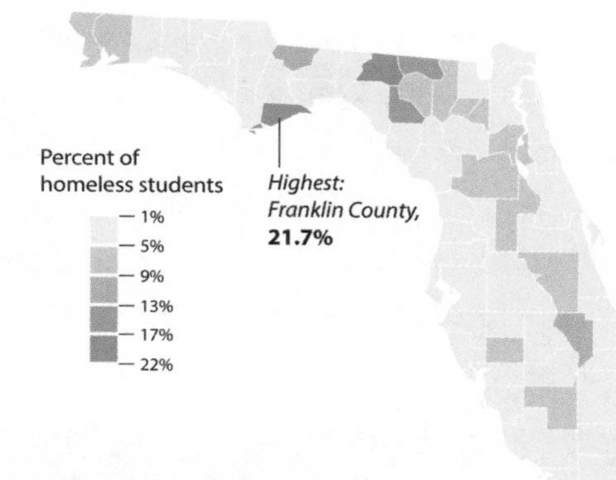

I was shocked. Did this mean that so many students were living on the streets? After all, that was my assumption as to what "homeless" means, but it's far from reality. According to Luís's story, Florida's public education system says that a student is homeless when he or she lacks "a fixed, regular, and adequate nighttime residence," or shares a home with people who aren't close relatives "due to loss of housing" or "economic hardship."

Therefore, most of these students don't live on the streets, but they don't have a stable home either. Even if the picture doesn't sound so dire now, it still is: lacking a permanent place, moving from home to home frequently, greatly lowers school performance, worsens behavior, and may have negative long-term consequences, as Luís's investigation showed. A debate about how to solve homelessness is critical, but to have it, we need to know exactly what the charts that plot it are measuring.

The internet and social media are powerful tools for creating, finding, and disseminating information. My social media streams brim with news and comments written or curated by journalists, statisticians, scientists, designers, and politicians, some of them friends, others unknown. We are all exposed to similarly endless spates of headlines, photos, and videos.

I love social media. It has helped me discover charts by many designers I'd never heard of and the writings of many authors I wouldn't have found otherwise. Thanks to social media, I can follow plenty of sources of good and dubious charts, such as FloorCharts, a Twitter account that collects bizarre visuals by members of Congress. Take a look at this one by Wyoming senator John Barrasso, which confuses percentage change and percentage *point* change: the increase between 39% and 89% isn't 50%, it's 50 *percentage points*—and also a 128% increase:

However, social media has its dark side. The core prompt of social media is to share, and to share quickly whatever captures our eyes, without paying much attention to it. This is why I promoted the heavy metal map mindlessly. It resonated with my existing likes and beliefs, so I re-sent it without proper reflection at first. Feeling guilty, I undid my action and verified the source before sending it out again.

The world will be a much better place if we all begin curbing our sharing impulses a bit more often. In the past, only professionals with access to publishing platforms—journalists and the owners of newspapers, magazines, or TV stations—controlled the information that the public received. Today, every single one of us is a curator of information, a role that implies certain responsibilities. One of them is to make sure, whenever we can, that whatever content we read and help spread looks legitimate, particularly if that content seems to confirm our most strongly rooted ideological beliefs and prejudices.

Sometimes, even lives may be at stake.

On the evening of June 17, 2015, a 21-year-old man called Dylann Roof entered Emanuel African Methodist Episcopal Church in Charleston, South Carolina. He asked for the pastor, Reverend Clementa Pinckney, a respected figure in the city and state and a legislator for nearly two decades.[2]

Pinckney took Roof to a Bible study session in the church basement, where the pastor discussed scripture with a small group of congregants. After a heated interchange, Roof pulled a handgun and murdered nine people. One victim pleaded with Roof to stop, and he replied, "No, you've raped our women and you are taking over the country. I have to do what I have to do." Roof's "you" meant "African Americans." Mother Emanuel, as the church is also known, is one of the oldest black congregations in the United States.

Roof was arrested and he became the first federal hate crime defendant to be sentenced to death.[3] In both his manifesto and confession, he explained the origins of his deep racial hatred. He referred to his quests for information about "black-on-white crime" on the internet.[4] His first source was the Council of Conservative Citizens (CCC), a racist organization that publishes articles with charts like this to make the point that black offenders target a disproportionate number of white victims *because they are white*:[5]

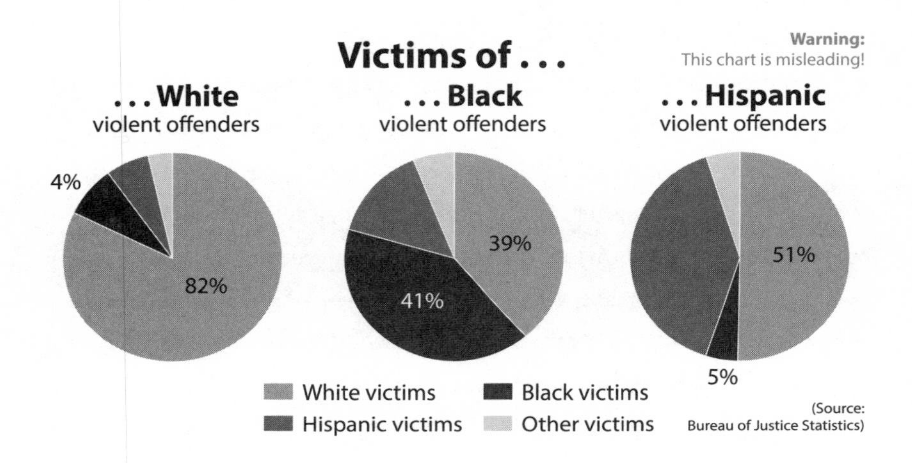

Human nature dictates that what we see is largely shaped by what we want to see, and Roof is no exception. His manifesto revealed a mind infected by racial grievances developed in childhood and adolescence and later solidified with data and charts twisted for political interest by an extremist organization. CCC's charts, designed by white supremacist Jared Taylor, were inspired by a confusing article by *National Review*'s Heather Mac Donald.[6] This case is a potent example of how crucial it is to check primary sources and read the fine print of how chart designers came up with the numbers they present.

Taylor's data came from the Victimization Survey by the Bureau of Justice Statistics, which can be easily found through a Google search. More specifically, it came from the table below. I added little arrows to indicate the direction in which you should read the numbers; the percentages inside the red boxes add up to 100% horizontally:

Distribution of violent victimizations, by race/Hispanic origin of victim and perceived race/Hispanic origin of offender, 2012-2013

Race of victim	Annual average number of victimizations	Total	Race of offender: White/a	Black/a	Hispanic	Other/a,b	Unknown
Total violence	6,484,507	100 % ↔	42.9	22.4	14.8	12.1	7.8
White/a	4,091,971	100 % ↔	56.0	13.7	11.9	10.6	7.8
Black/a	955,800	100 % ↔	10.4	62.2	4.7	15.0	7.7
Hispanic	995,996	100 % ↔	21.7	21.2	38.6	11.6	6.9
Other/a,b	440,741	100 % ↔	40.3	19.3	10.6	20.3	9.5

a/Excludes persons of Hispanic or Latino origin
b/Includes American Indian and Alaskan Native, Asian, Hawaiian, other Pacific Islander, and persons of two or more races

(Source: Bureau of Justice Statistics, National Crime Victimization Survey)

The table displays violent crimes with the exception of murders. Notice that "white" and "black" exclude Hispanic and Latino whites and blacks. "Hispanic" means anyone of Hispanic or Latino origin, regardless of color or race.

Understanding the difference between the numbers on the table and the figures Taylor crafted is tricky. Let's begin by verbalizing what the table tells us. Trust me, even I would have a hard time wrapping my head around the numbers if I didn't explain them to myself:

- There were nearly six and a half million victims of violent crimes—excluding murders—in both 2012 and 2013.
- Of those, a bit more than four million victims (63% of the total victims) were whites and nearly one million (15% of the total victims) were blacks. The rest belong to other races or ethnicities.
- Focus now on the "White" row: 56% of white victims were attacked by white offenders; 13.7% were attacked by black offenders.
- Let's move to the "Black" row: 10.4% of black victims were attacked by white offenders; 62.2% were attacked by black offenders.

What the table says—and what is true about this data—is the following:

The percentage of victims who are non-Hispanic whites and blacks is very close to their presence in the U.S. population: 63% of the victims were non-Hispanic white, and 61% of people in the United States were non-Hispanic white, according to the Census Bureau (and more than 70% if we count whites who are also Hispanic or Latino); 15% of the victims were black, and 13% of the U.S. population is African American.

When people become victims of crimes, it's more likely than not that their aggressors are of their same demographic. Let's do a chart with the same format as Taylor's but displaying the numbers as they should be:

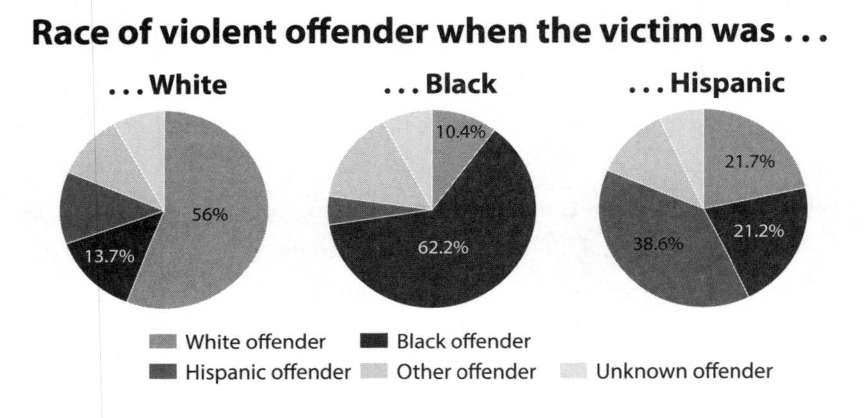

How is it that Taylor's figures are so different from those reported by the Bureau of Justice Statistics? The reason is that Taylor played an arithmetic sleight of hand to serve a preconceived message intended to spur racial animus. In his own words, "When whites commit violence they choose fellow whites [most of the time] and almost never attack blacks. Blacks attack whites almost as often as they attack blacks."

To come up with his numbers, Taylor first transformed the percentages on the Bureau of Justice Statistics table into the number of victims. For instance, if the table shows that there were 4 million white victims and that 56% of them where attacked by white offenders, then roughly 2.3 million white victims were attacked by other whites.

The first table Taylor designed probably looked like the one on the top of the following page.

Race/ethnicity of victims	Annual average number of victimizations	White offender	Black offender	Hispanic offender	Other offender	Unknown offender
Total	6,484,507	2,781,854	1,452,530	959,707	784,625	505,792
White	4,091,971	2,291,504	560,600	486,945	433,749	319,174
Black	955,800	99,403	594,508	44,923	143,370	73,597
Hispanic	995,996	216,131	211,151	384,454	115,536	68,724
Other	440,741	177,619	85,063	46,719	89,470	41,870

Taylor then read this table column by column, from top to bottom, and used the figures on the "Total" row as denominators to transform all other numbers into percentages. For instance, take a look at the "Black Offender" column. The total number of victims of black offenders was 1,452,530. Out of those, 560,600 were white, which is 38.6% of 1,425,530. Here are the results that Taylor got and that he represented in his pie charts:

Race/ethnicity of victims	Annual average number of victimizations	White offender	Black offender	Hispanic offender	Other offender	Unknown offender
Total	6,484,507	2,781,854	1,452,530	959,707	784,625	505,792
White	63.1 %	82.4 %	38.6 %	50.7 %	55.3 %	63.1 %
Black	14.7 %	3.6 %	40.9 %	4.7 %	18.3 %	14.6 %
Hispanic	15.4 %	7.8 %	14.5 %	40.1 %	14.7 %	13.6 %
Other	6.8 %	6.4 %	5.9 %	4.9 %	11.4 %	8.3 %

(Notice that the only small discrepancy between my calculation and Taylor's is that I have 82.4% of white victims of white offenders and he had 82.9%.)

Arithmetically speaking, these percentages may be correct, but arithmetic isn't the main factor that makes a number meaningful. Numbers must always be interpreted in context. Taylor made at least four dubious assumptions.

First, he ignored the racial makeup of the United States. According to the Census Bureau, as of 2016 the U.S. population was about 73% white (including Hispanic and Latino whites) and about 13% black. Based on this fact, my University of Miami graduate student and data analyst Alyssa Fowers did the following back-of-the-napkin calculation for me:

> If a hypothetical (and very active!) white offender committed crimes against people of his or her own race half the time, and then committed the other half of their crimes randomly among the entire popula-

> tion, they would commit 86.5% of their crimes against white people and 6.5% against black people.
>
> Meanwhile, if a black offender did the exact same thing—half against people of their own race, half randomly against the entire population—they would commit only 56.5% of their crimes against black people and 36.5% of their crimes against white people. That makes it look like the black offender is intentionally targeting white victims more than the white offender is targeting black victims, when in fact there are just many more potential white victims and many fewer potential black victims because of the population makeup of the United States.

The second dubious assumption Taylor made is that his aggregation is better than the one by the Bureau of Justice Statistics. The opposite is true because of the nature of these violent crimes: offenders often victimize people who are like them and live nearby. Many violent crimes are the product of domestic violence, for instance. The Bureau of Justice Statistics explains that "the percentage of intraracial victimization [is] higher than the percentage of interracial victimization for all types of violent crime except robbery." Robbery may be the exception because if you're a robber, you'll try to assault people who live in wealthier areas than you do.

This last fact is related to Taylor's third wrong assumption: that offenders "choose" their victims because of their race, that blacks "choose" whites as victims more often than whites "choose" blacks. Unless is a crime is premeditated, criminals don't choose their victims at all, much less by their race. In the most common kinds of violent crimes, offenders attack victims because they are furious with them (domestic violence) or because they expect to get something valuable from them (robbery). Do black offenders rob white victims? Sure. But that's not a racially motivated crime.

The fourth assumption is the most relevant one. Taylor wanted readers to believe that truly racially motivated crimes—hate crimes—aren't counted.

They are, in fact, counted, and they would have been more appropriate for his discussion, although these figures are quite inconvenient for him: In 2013, law enforcement agencies reported that 3,407 hate-crime offenses were racially motivated. Of these, 66.4% were motivated by antiblack bias, and 21.4% stemmed from antiwhite bias.[7]

These are the numbers that should have appeared in Taylor's charts. As George Mason University professor David A. Schum wrote in his book *The Evidential Foundations of Probabilistic Reasoning*,[8] data "becomes evidence in a particular inference when its relevance to this inference has been established." The fact that many offenders are black and many victims are white is not evidence for the inference that offenders choose their victims or that this potential choice is racially motivated.

It's hard not to wonder what would have happened if Dylann Roof had found numbers like these instead of those fudged by the Council of Conservative Citizens. Would he have changed his racist beliefs? It seems unlikely to me, but at least those beliefs wouldn't have been further reinforced. Dubious arithmetic and charts may have lethal consequences.

Economist Ronald Coase once said that if you torture data long enough, it'll always confess to anything.[9] This is a mantra that tricksters have internalized and apply with abandon. As the charts that confirmed Dylann Roof's racist beliefs demonstrate, the same numbers can convey two opposite messages, depending on how they are manipulated.

Imagine that I run a company with 30 employees, and in the annual report that I send to shareholders, I mention that I care about equality and employ the same number of men and women. In the document, I also celebrate that three-fifths of my female employees have higher salaries than male employees of the same rank, to compensate for the fact that women in the workforce tend to make less money than men. Am I lying? You won't know unless I disclose all the data as a table:

Female employees

Employee	Salary ($)	Employee	Salary ($)
Manager	150,000	Regular worker	45,000
Manager	130,000	Regular worker	42,000
Manager	115,000	Regular worker	40,000
Supervisor	76,000	Regular worker	38,000
Supervisor	74,500	Regular worker	36,000
Supervisor	72,000	Regular worker	35,250
Regular worker	70,000	Intern	15,000
		Intern	15,000

Woman makes more than man in the same position

Male employees

Employee	Salary ($)	Employee	Salary ($)
Manager	162,000	Regular worker	44,750
Manager	138,500	Regular worker	41,000
Manager	125,000	Regular worker	39,500
Supervisor	80,000	Regular worker	37,000
Supervisor	76,000	Regular worker	35,500
Supervisor	73,000	Regular worker	35,000
Regular worker	68,500	Intern	14,000
		Intern	14,000

Man makes more than woman in the same position

I may not have told you a complete lie, but I'm not disclosing the whole truth, either. A majority of my women employees have higher salaries than the men, but I concealed the fact that, on average, men in my company make more than women (the means are $65,583 and $63,583), because managerial salaries are so unequal. Both ways of measuring equality are relevant if I want to provide a truthful picture of my company.

This is a fictional example, but similar ones in the news media abound. On February 22, 2018, BBC News wrote, "Women earn up to 43% less at Barclays—Female employees earn up to 43.5% less at Barclays than men, according to gender pay gap figures it has submitted to the government."[10] This isn't a lie, either. The pay gap at Barclays Bank is indeed large. However, as data analyst Jeffrey Shaffer pointed out,[11] that 43.5% difference doesn't tell the entire story. We need to take a look at charts like this because they reveal an angle we may have overlooked:

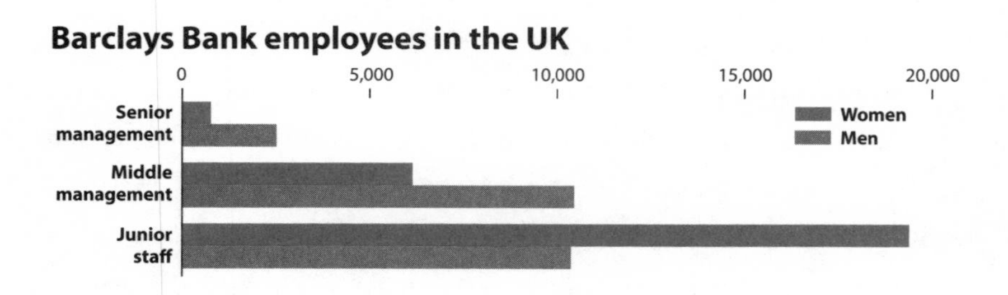

Barclays does have an equality problem, but it isn't a pay gap at the same levels in the hierarchy—men and women in similar positions earn nearly identical salaries, according to a report by the bank. Barclays' challenge is that women employees are mostly in junior positions and managers are mostly men, so the key to solving the problem may be its promotion policies. Jes Staley, CEO of the bank, said, "Although female representation is growing at Barclays, we still have high proportions of women in more junior, lower paid roles and high proportions of men in senior, highly paid roles."

Numbers can always yield multiple interpretations, and they may be approached from varied angles. We journalists don't vary our approaches more often because many of us are sloppy, innumerate, or simply forced to publish stories at a quick pace. That's why chart readers must remain vigilant. Even the most honest chart creator makes mistakes. I know this because I've made most of the mistakes I call out in this book—even though I didn't wish to lie on purpose!

On July 19, 2016, the news website Vox published a story with the title "America's Health Care Prices Are Out of Control. These 11 Charts Prove It."[12]

Here's a mantra I like to repeat in classes and talks: **Charts alone rarely prove anything.** They can be powerful and persuasive components of arguments and discussions, but they are usually worthless on their own. The charts Vox mentioned in the title look like this:

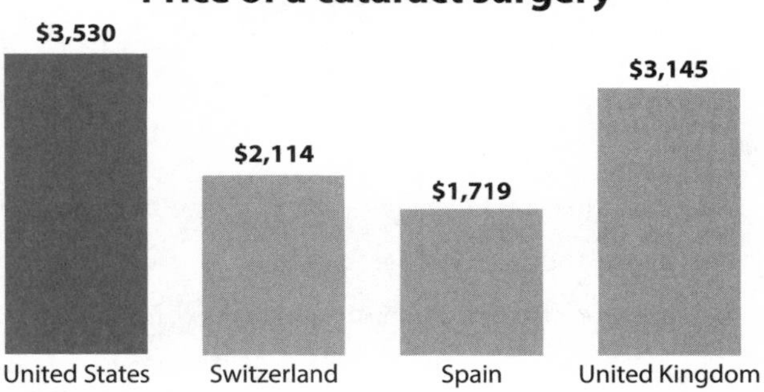

Vox's story is the type of content I'd feel tempted to promote in social media, as it confirms something I already believe: I was born in Spain, where health care is largely public and paid for through taxes, as in many other western European countries. Of course I believe that health care prices in the United States are "out of control"—I've suffered them myself!

However, how out of control are they, really? Vox's charts made my internal baloney alarm ring because the story didn't mention whether the prices reported were adjusted by purchasing power parity (PPP). This is a method to compare prices in different places taking into account cost of living and inflation. It's based on calculating what quantity of each country's currency you need if you want to buy the same basket of goods. Having lived in many different places, I can assure you that $1,000 is a lot of money in some of them and not so much in others.

I thought about PPP also because many members of my family in Spain worked in health care: my father was a medical doctor before retiring, and so was my uncle; my father's sister was a nurse; my mother used to be head of nursing in a large hospital; my paternal grandfather was a nurse. I'm familiar with the salaries they used to make. They were less than half of what they would have made if they had moved to the United States to work in the same kinds of jobs, which is the proportion I saw in some of Vox's charts.

Curious about where the data had come from and about whether prices had been adjusted to make them truly comparable, I did some digging online. Vox's source, mentioned in the story, was a report by the International Federation of Health Plans, or IFHP (http://www.ifhp.com/), based in London. Its membership is 70 health organizations and insurers in 25 countries.[13]

The report contains an Overview page explaining the methodology employed to estimate the average price of several health care procedures and drugs in different countries. It begins with this line:

Prices for each country were submitted by participating federation member plans.

This means that the survey didn't average the prices of *all* health plan providers in all countries but rather those from a sample. This isn't wrong in principle. When estimating anything—the average weight of U.S. citizens, for instance—it's unlikely that you'll be able to measure every single individual. It's more realistic to draw a large *random sample* and then average the measurements taken from it. In this case, a proper sample would consist of a small set of randomly picked health plan providers per country. All providers should be equally likely to be selected.

If random sampling is rigorously conducted,[14] it's likely that the average you compute based on it will be close to the average of the population the sample was drawn from. A statistician would say that a carefully selected random sample is "representative" of the population it was drawn from. The average of the sample won't be identical to the population average, but it'll be similar. That's why statistical estimates are often accompanied by an indication of the uncertainty that surrounds them, such as the famous "margin of error."

But the sample that the IFHP used isn't random. It's *self-selected*. It averages the prices reported by organizations that chose to be members of the IFHP. Self-selected samples are risky because there's no way to assess whether the statistics calculated from them correspond to the populations they stand for.

I bet you're familiar with another egregious example of self-selected sampling: polls on websites and social media. Imagine that the leftist magazine *The Nation* asked on social media whether respondents approve of a Republican president. The results they'd get would be 95% disapprove and 5% approve, which would be unsurprising, considering that readers of *The Nation* are likely progressives and liberals. The results of such a poll would be the opposite if conducted by Fox News.

It gets worse. Next on the IFHP report's Overview page we find this paragraph:

> Prices for the United States were derived from over 370 million medical claims and over 170 million pharmacy claims that reflect prices negotiated and paid to health care providers.

But prices for other countries . . .

> . . . are from the private sector, with data provided by one private health plan in each country.

This is problematic because we don't know whether that single private health plan in each country is representative of *all* plans in that same country. It may be that the price for a cataract surgery reported by the single Spanish health plan provider in the sample coincides with the average price of a cataract surgery in Spain. But it may also be that this specific plan is either much more expensive than the country average or much cheaper. We just don't know! And neither does the IFHP, as they openly acknowledge in the last line of the overview:

> A single plan's prices may not be representative of prices paid by other plans in that market.

Well, yes, indeed. Surreptitiously, this sentence tells us, "If you use our data, please warn your readers of their limitations!" Why didn't Vox explain the data's many limitations in the story, so readers could have taken the numbers with a grain—or a mountain—of salt? I don't know, but I can make a guess, as I've done plenty of flawed charts and stories myself: a crushing majority of journalists are well intentioned, but we're also busy, rushed, and, in cases like mine, very absentminded. We screw up more often than we like to admit.

I don't think that this should lead us to mistrust all news media, as I'll explain at the end of this chapter, but it ought to make us think carefully of the sources from which we obtain information and always apply common-sense reasoning rules such as Carl Sagan's famous phrase "Extraordinary claims require extraordinary evidence."

Here's an example of an extraordinary claim: People in states that lean Democratic consume more porn than people in Republican states. The exception is Kansas, according to data from the popular website Pornhub.[15] Jayhawkers, on average, consume a disproportionate amount of online porn:

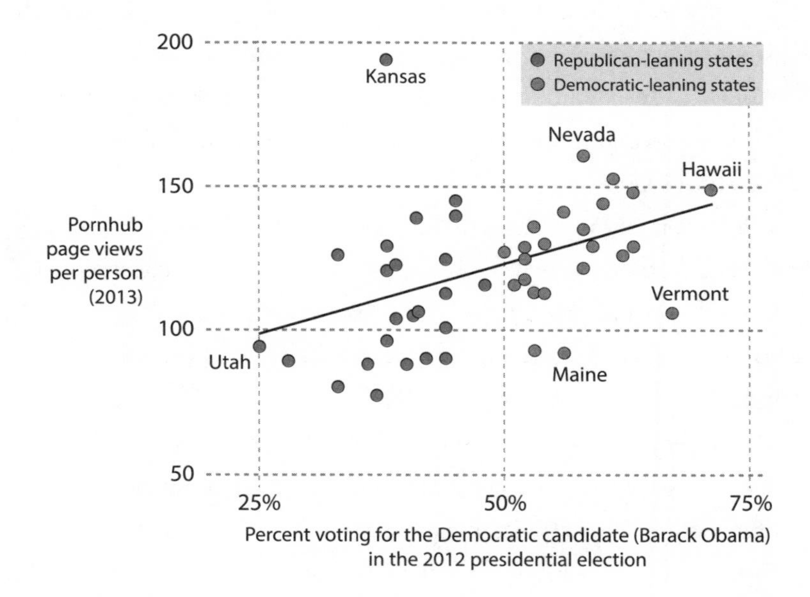

Oh Kansans, how naughty of thee! You watch much more porn (194 pages per person) than those northeastern heathen liberals from Maine (92) or Vermont (106).

Except that you don't. To explain why, I first need to show you where the geographic center of the contiguous United States is:

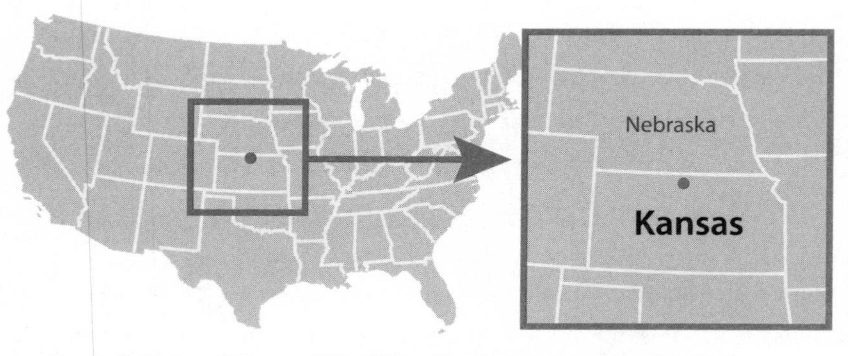

• Geographic center of the contiguous United States

The preceding scatter plot is based on one designed by reporter Christopher Ingraham for his personal blog about politics and economics, WonkViz. Ingraham's scatter plot and the Pornhub data it used got picked up by several news publications. They all had to issue corrections to their stories later on.

The data and the inferences made from it are problematic. First, we don't know if Pornhub views are a good reflection of overall porn consumption—it may be that people in different states use other sources. Moreover, the reason why porn consumption per capita in Kansas looks so high is a quirky glitch in the data. Unless you use tools such as a virtual private network (VPN), the people who run websites and search engines can pinpoint your location thanks to your internet protocol (IP) address, a unique numerical identifier assigned to your internet connection. For example, if I visit Pornhub from my home in Florida, the Pornhub data folks will know roughly where I am.

However, I do use a VPN that redirects my internet traffic to servers located in different parts of the world. Right now, my VPN server is in

Santa Clara, California, even if I'm comfortably writing from my sunny backyard in Florida. If I'm added to Pornhub's databases, I may be added to either "Santa Clara, CA," or, as they'd probably know I'm using a VPN, to "Location Not Determined." However, this is not what happened in this case. If I can't be located, I won't be removed from the data, but I'll be automatically assigned to the center of the contiguous United States, making me a Kansan. Here's Ingraham pointing out the erroneous message of the scatter plot:

> Kansas' strong showing is likely an artifact of geolocation—when a U.S. site visitor's exact location can't be determined by the server, they are placed at the center of the country—in this case, Kansas. So what you're seeing here is likely a case of Kansas getting blamed (taking credit for?) anonymous Americans' porn searches.[14]

When journalists and news organizations acknowledge errors and issue corrections, as Ingraham did, it's a sign that they are trustworthy.

Another sign of trustworthiness is whether a journalist explores the data from multiple angles and consults different sources. Out of curiosity, I casually delved into the literature about the relationship between porn consumption patterns and political leanings—there's such a thing—and discovered a paper in the *Journal of Economic Perspectives* titled "Red Light States: Who Buys Online Adult Entertainment?" by Benjamin Edelman, a professor of business administration at Harvard.[16]

If Pornhub showed that, on average, people in states that leaned liberal in 2012 consumed more porn from its website, this paper reveals the opposite pattern: it's red states that consume more adult entertainment. Here's a quick chart I designed based on Edelman's data (note: Edelman didn't include all states, and the inverse association between the variables is quite weak):

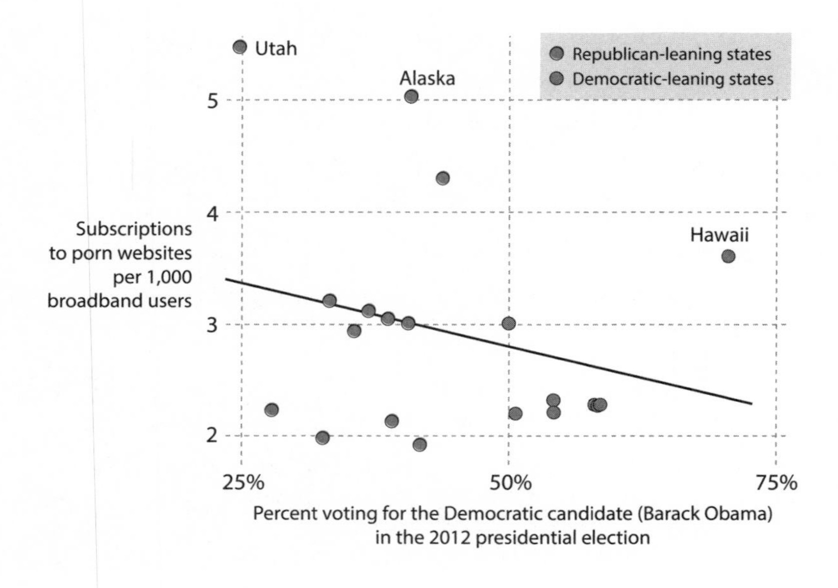

Some outliers in this case are Utah, Alaska, and Hawaii. The key feature to pay attention to when comparing this chart with the one before is the label on the vertical axis: on Ingraham's chart, it was Pornhub pages per person; here, it's subscriptions to porn websites per 1,000 broadband users.

To read a chart correctly, we must first identify what exactly is being measured, as this may radically change the messages a chart contains. Just an example: I cannot claim *from my latest chart alone* that Alaska, Utah, or Hawaii consume more porn; it might well be that people in those states watch less porn but get it more often from paid-for websites, rather than from freemium ones such as Pornhub. Also, as we'll see in chapter 6, we can't claim from a chart displaying state figures that each *individual* from those states consumes more or less porn.

Being an attentive chart reader means being a critical reader of data. It also requires you to develop a sense of what sources you can trust. Both these goals are beyond the scope of this book, but I'd like to offer some tips.

There are books that will help you to become better at assessing the numbers we see in the media. I personally recommend Charles Wheelan's *Naked Statistics*, Ben Goldacre's *Bad Science*, and Jordan Ellenberg's *How Not to Be Wrong*. These books alone may help you avoid many of the most common mistakes we all make when dealing with everyday statistics. They barely refer to charts, but they are excellent books from which to learn some elementary data-reasoning skills.

To be a good media consumer, I suggest the Fact-Checking Day website (https://factcheckingday.com). It was created by the Poynter Institute, a nonprofit school dedicated to teaching information literacy and journalism. The website offers a list of features that may help you determine the trustworthiness of a chart, a news story, or an entire website or publication.

On a related note, everyone who has an online presence today is a publisher, a role that used to be played just by journalists, news organizations, and other media institutions. Some of us publish for small groups of people—our friends and family—while others have big followings: my own Twitter following includes colleagues, acquaintances, and complete strangers. No matter the number of followers we have, we can all potentially reach thousands, if not millions, of people. This fact begets moral responsibility. We need to stop sharing charts and news stories mindlessly. We all have the civic duty to avoid spreading charts and stories that may be misleading. We must contribute to a healthier informational environment.

Let me share my own set of principles for spreading information so that you can create your own. Here's my process: Whenever I see a chart, I read it carefully and take a look at who published it. If I have time, I visit the primary source of the data, as I did with the heavy metal map and Vox's story about health care prices. Doing this for a few minutes before sharing a chart doesn't guarantee that I won't promote faulty content every now and then, but it reduces the likelihood of that happening.

If I have doubts about a chart or story based on data, I don't share it. Rather, I ask people I trust and who know more about the subject. For example, this book and all the charts it contains were read by a few friends with PhDs in data-related fields before being published. If I can't asses the quality of a chart on my own, I rely on people like them to help me. You don't need to befriend any eggheads to do this, by the way; your kids' math or science teacher will do just fine.

If I can explain why I believe a chart is wrong or can be improved upon, I publish that explanation in social media or in my personal website along with the chart. Then I call the author's attention to it and do my best to be constructive in these critiques (unless I'm certain of ill intentions). We all make mistakes and we can learn from each other.

It's unrealistic to expect that all of us can verify the data of *all* charts we see every day. We often don't have time, and we may lack the knowledge to do so. We need to rely on trust. How do we decide whether a source is trustworthy?

Here's a set of very personal rules of thumb based on my experience and what I know about journalism, science, and the shortcomings of the human brain. In no particular order, they are:

- Don't trust any chart built or shared by a source you're not familiar with—until you can vet either the chart or the source, or both.
- Don't trust chart authors and publishers who don't mention the sources of their data or who don't link directly to them. Transparency is another sign of appropriate standards.
- Try a varied media diet, and not only for charts. No matter your ideological standpoint, seek out information from people and publications on the right, in the center, and on the left.

- Expose yourself to sources you disagree with, and assume good faith on their part. I'm convinced that most people don't want to lie or mislead on purpose and that we all loathe being lied to.
- Don't assume ill intentions when haste, sloppiness, or ignorance is the more likely explanation for a bad chart.
- Needless to say, trust has its limits. If you begin spotting a pattern of misdeed in a source on your list, erase it.
- Follow only sources that issue corrections when they ought to and that do it visibly. Corrections are another sign of high civic or professional standards. To co-opt a common saying: To err is human, to correct is divine. If a source you follow doesn't issue corrections systematically after being proved wrong, drop it.
- Some people think that all journalists have agendas. This happens in part because many relate journalism to raging pundits on TV and talk radio. Some of those are journalists, but many aren't. They are either entertainers, public relations specialists, or partisan operatives.
- All journalists have political views. Who doesn't? But most try to curb them, and they do their best to convey, as famed Watergate reporter Carl Bernstein likes to say, "the best obtainable version of the truth."[17]
- This obtainable version may not be *the* truth, but good journalism is a bit like good science. Science doesn't discover truth. What science does well is provide increasingly better approximate explanations of what the truth may be, according to available evidence. If that evidence changes, the explanations—either journalistic or scientific—should change accordingly. Beware those who never change their views despite acknowledging that their previous views were driven by incomplete or faulty data.
- Avoid very partisan sources. They don't produce information but pollution.
- Telling the difference between a merely partisan source—there are reliable ones all over the ideological spectrum—and a hyperpartisan one

can be tricky. It will require some time and effort on your part, but there is a very good clue you can begin with: the tone of the source's messages, including whether that source employs ideologically loaded, bombastic, or aggressive language. If it does, stop paying attention to it, even if it's just for entertainment.

- Hyperpartisan sources, particularly those you agree with, are similar to candy: A bit of it from time to time is fine and fun. A lot of it regularly is unhealthy. Nourish your mind instead, and push it to be exercised and challenged rather than coddled. Otherwise, it'll wither.
- The more ideologically aligned you are with a publication, the more you should force yourself to read whatever it publishes with a critical eye. We humans find comfort in charts and stories that corroborate what we already believe and react negatively against those that refute it.
- Expertise matters, but it's also specific. When it comes to arguing over a chart about immigration, your judgment as a layperson is as valid as that of a mechanical engineer or someone with a PhD in physics or philosophy. And your opinion is less likely to be accurate than are the ones expressed by statisticians, social scientists, or attorneys who specialize in immigration. Embrace intellectual modesty.
- It's become fashionable to bash experts, but healthy skepticism can easily go too far and become nihilism, particularly if for emotional or ideological reasons you don't like what certain experts say.[18]
- It's easy to be overly critical of charts that depict realities we'd rather not learn about. It's much harder to read those charts, to assume that their creators likely acted in good faith, and then to coolly assess whether what the charts show has merit. Don't immediately snap to judgment against a chart just because you dislike its designers or their ideology.

Finally, remember that one reason charts lie is that we are prone to lying to ourselves. This is a core lesson of this book, as I'll explain at length in the conclusion.

Chapter 4

Charts That Lie by Displaying Insufficient Data

Peddlers of visual rubbish know that cherry-picking data is an effective way to deceive people. Choose your numbers carefully according to the point you want to make, then discard whatever may refute it, and you'll be able to craft a nice chart to serve your needs.

An alternative would be to do the opposite: Rather than display a small amount of mischievously selected data, throw as much as possible into the chart to overwhelm the mental bandwidth of those you want to persuade. If you don't wish anyone to notice a specific tree, show the entire forest at once.

On December 18, 2017, my day was disrupted by a terrible chart tweeted by the White House. It's a core tenet of mine that if you want to have cordial and reasonable discussions about thorny issues, you need to use good evidence. The chart, which you can see on the following page, doesn't fit the bill.

Curious about the chart, I followed the link and saw that it was part of a series against family-based migration. Some graphics pointed out that 70% of U.S. migration in the last decade is family based (people bringing relatives over), which amounts to 9.3 million immigrants.[1]

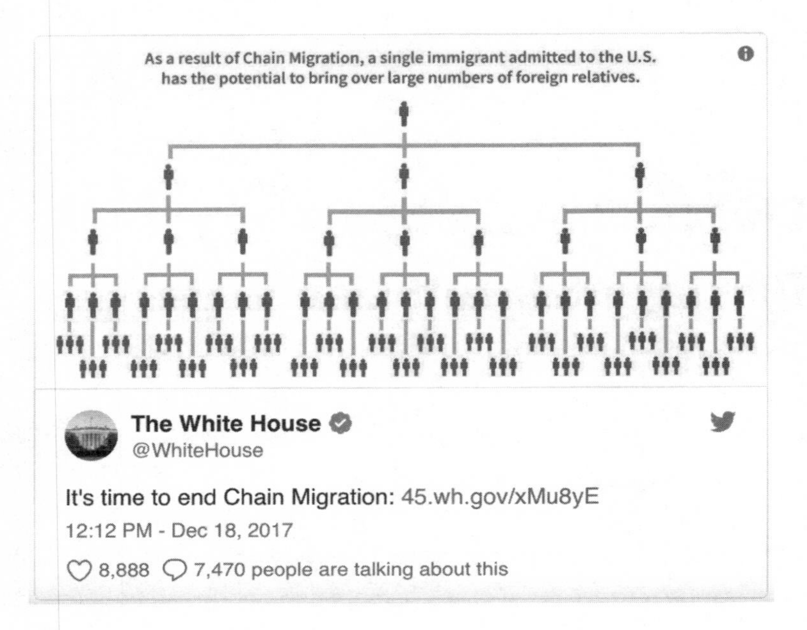

I don't have strong opinions either in favor of or against family-based migration. I've heard good arguments on both sides. On one hand, letting immigrants sponsor relatives beyond their closest circle isn't just a humane thing to do, it also may have psychological and societal benefits; extensive and strong family networks provide protection, security, and stability. On the other hand, it might be a good idea to strengthen merit-based migration and increase the number of highly skilled individuals, in exchange for limiting other kinds of migration.

What I do have strong opinions about are propaganda and misleading charts. First, notice the loaded language, something I warned against in the previous chapter: "chain migration" is a term that has been used widely in the past, but "family-based migration" is much more neutral.

Here's how the White House describes people moving to the United States: "In the last decade, the U.S. has permanently resettled 9.3 million immigrants solely on the basis of familial ties." Resettled? I'm an immigrant myself. I was born in Spain, and my wife and children were born in Brazil. We weren't "resettled by the U.S." We *moved* here. And if we ever sponsor relatives, they wouldn't be "resettled" either; they'd also be moving here voluntarily.

The language the White House used is intended to bias your opinion before you even look at the data. This is a trick that is based on sound science. We human beings form judgments based on quick emotional responses and then use whatever evidence we find to support those opinions, rather than rationally weigh data and then make up our minds. As psychologist Michael Shermer pointed out in his book *The Believing Brain*, forming beliefs is easy; changing them is hard work.[2] If I use charged language, I may surreptitiously trigger an emotional response in my audience that will bias their understanding of the charts.

Moreover, the rhetorical style of the White House charts is also tailored to spur emotions. Just see *how many* new immigrants a single newcomer "has the potential" to spawn! They look like bacteria, vermin, or cockroaches tripling in every generation, a metaphor that has dark historical precedents. The chart the White House tweeted looks eerily similar to charts favored by racists and eugenics proponents. Here's a German 1930s chart depicting the "danger" of letting the "inferior" races multiply without control:

Charts may lie because they are based on faulty data, but they may also fail because they seem to provide insights while containing no data whatsoever. The White House chart is an example of this.

Who is that prototypical migrant who will end up bringing dozens of relatives? We don't know. Is that person representative of all migrants? Not at all. Here's how I know: I arrived in the United States in 2012 with an H-1B visa, reserved for people with specialized skills, sponsored my wife and two kids, and later got a Green Card, becoming a permanent resident. You can think of me as the guy on top of the chart and my family as being on the second step.

So far, so good: me on top, three relatives below me on the chart. The chart is accurate up to that point, although the White House forgot to mention that a majority of the family-based visas assigned every year are for families like mine, made of two spouses and their unmarried children. I don't think that even the staunchest anti-immigration activist would want to eliminate this policy, although I may be wrong.

But what happens in the lower levels of the chart, when the migrants on the second step start bringing three other relatives each? Well, it's not that easy: if my wife wanted to bring her mother and siblings, she'd need to sponsor them as *non-immediate* relatives, a category that excludes extended family such as uncles or cousins. Moreover, to apply for sponsorship in certain visa categories, my wife would need first to become a citizen, so she wouldn't be a "migrant" anymore.

The number of family-based visas is capped at 480,000 a year. According to the National Immigration Forum, there are no limits on visas for immediate relatives, but this number is subtracted from the 480,000 limit, and therefore the number of visas going to non-immediate relatives is much lower. This means that you can't bring anyone you want and, moreover, that bringing a non-immediate relative may take years instead, as the number of visas that can go to any one country is limited.

Politically charged topics often yield the best—or worst, depending on your perspective—examples of faulty charts and data. For example, in September 2017, a headline on Breitbart News claimed that "2,139 DACA Recipients Convicted or Accused of Crimes against Americans."[3]

DACA stands for Deferred Action for Childhood Arrivals. It's a policy announced by President Barack Obama in 2012 that protects from deportation some people who were brought to the United States illegally as children and gives them work permits. DACA has many critics who say that it's a policy created by the executive branch and that it should have been instead discussed in Congress. Some voices I deem reasonable even say it's unconstitutional,[4] and President Trump ended it in September 2017.

This debate is beside the point, though. Let's focus on the fact that worthy debates can be hindered by flawed graphics. I've designed one based on Breitbart's data point, trying to match the strident rhetorical style of the article:

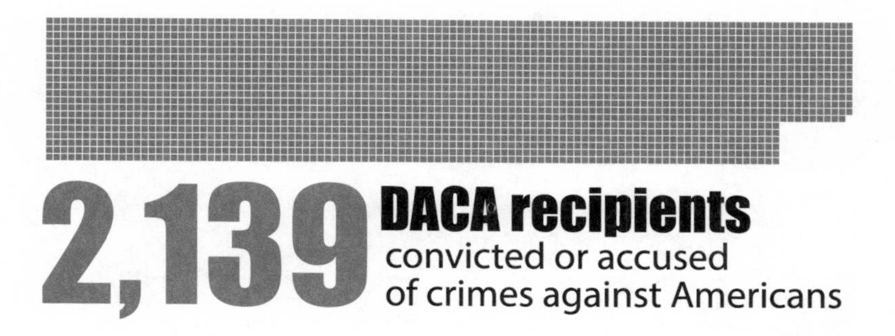

The first paragraph of the story says:

> As Attorney General Jeff Sessions announced the end of the Obama-created Deferred Action for Childhood Arrivals (DACA), from which more than 800,000 un-vetted young illegal aliens have been given protected status and work permits, the number of them who are convicted criminals, gang members, or suspects in crimes remains staggering.

It's a staggering number indeed, that 2,139.

Not staggeringly high, however, but rather staggeringly *low*. According to the article, there are more than 800,000 DACA recipients. If that is true, the proportion of them who lost their status for an accusation of being part of a gang or for a criminal conviction is very small.

Let's do some simple math: If you divide 2,139 by the denominator 800,000, you get roughly 0.003. If you multiply that by 100, you get a percentage: 0.3%. If we multiply by 1,000 instead, we get another rate: out of every 1,000 DACA recipients, 3 lost their status because of their misdeeds.

The number looks even lower if we compare it to something else, which is what we always ought to do. Single figures are meaningless if they aren't put in context. We could compare that 3 out of 1,000 DACA recipients to a similar proportion of the entire U.S. population. A 2016 study estimates that in 2010 "ex-felons comprise 6.4 percent of the voting age population."[5] That is, 64 over 1,000 people:

Out of every 1,000 DACA recipients . . .

3 have had their temporary protected status revoked due to "a felony criminal conviction; a significant misdemeanor conviction; multiple misdemeanor convictions; gang affiliation; or arrest for any crime in which there is deemed to be a public safety concern."

Out of every 1,000 voting-age people who live in the U.S. . . .

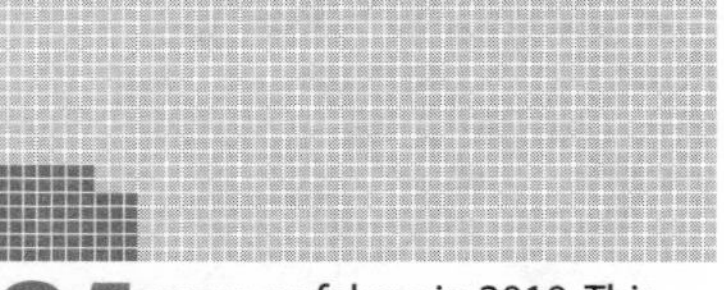

64 were ex-felons in 2010. This doesn't include people convicted of misdemeanors.

(Source: Shannon, Sarah K.S., et. al. "The growth, scope, and spatial distribution of people with felony records in the United States, 1948 to 2010." Demography 54(5)(2017): 1795-1818)

My comparison is much more informative than Breitbart's simplistic article but also imperfect for several reasons: First, this is just one estimate calculated by several scholars (I haven't, however, managed to find estimates that are significantly smaller). Second, the estimate for the entire U.S.

population includes people of all ages. To make a more accurate comparison, we'd need to calculate rates for people who are at most in their thirties, as all DACA recipients are around that age or younger.

Finally, those 3 out of every 1,000 DACA recipients lost their protections not only for felonies but also for misdemeanors and other minor offenses, and the estimate for the entirety of the U.S. population counts only felonies. As the academics who wrote the 2016 study explained:

> A felony is a broad categorization, encompassing everything from marijuana possession to homicide. Historically "felony" has been used to distinguish certain "high crimes" or "grave offenses" from less serious, misdemeanor offenses. In the United States, felonies are typically punishable by more than one year in prison, while misdemeanors garner less severe sanctions such as shorter jail sentences, fines, or both.

It may be that if we counted only DACA recipients expelled because of felonies, their number on the chart above would be even lower. We won't know for sure, however, unless there are more studies.

My first chart based on Breitbart's data point is an example of a chart that lies because it displays an inadequate amount of data, too little in this case. It also belongs to a subcategory: those charts that, sometimes to help push an agenda, cherry-pick the data and display counts when they should show rates instead, or vice versa.

No chart can ever capture reality in all its richness. However, a chart can be made worse or better depending on its ability to strike a balance between oversimplifying that reality and obscuring it with too much detail. In November 2017, former Speaker of the House Paul Ryan went to social media to promote the Tax Cuts and Jobs Act, a measure passed that same month. He used a graphic like this:

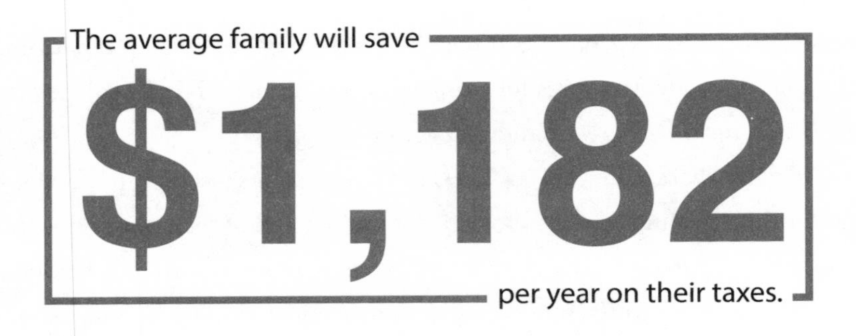

Regardless of what you think of the 2017 tax cut, this graphic is an oversimplification. Averages alone don't tell you much. How many families in the United States are "average" or close to "average"? A large majority? If I didn't know better, that's what I'd think if I trusted Ryan's figure.

According to the Census Bureau, the median household in the U.S. made $60,000 at the moment of writing this page.[6] (Note: Family income may not be the same as household income; a household is one or more people living in the same housing unit, but not all households contain families, which are several people who are related by birth, adoption, or marriage. However, the distributions of family incomes and of household incomes have similar shapes.)

Let's design a fictional chart that shows that a large majority of households had incomes close to $60,000.

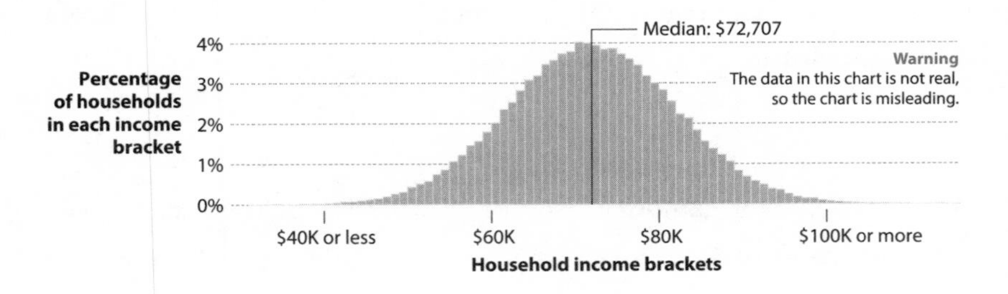

This kind of chart is called a histogram, and it's used to show frequencies and distributions, in this case, a hypothetical—and false—distribution of U.S. households depending on their income. In this histogram, the height of the bars is proportional to the percentage of households in each income bracket. The taller the bar, the more households there are at that particular income level. Also, if we stacked all the bars on top of each other, they'd add up to 100%.

On my fictional chart, the tallest bars are in the middle, close to the median. In fact, a vast majority of households have incomes between $40,000 and $80,000. The real distribution of household income in the United States is very different, though. Here it is:

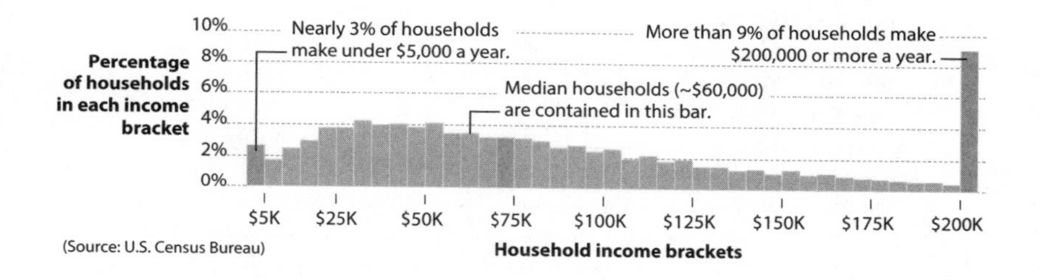

The range of household incomes in the U.S. is very wide, spanning from less than $5,000 a year to many millions of dollars. Income distribution is so skewed that we can't even show it in its entirety on the chart; we need to group wealthy households inside the "$200,000 or more a year" bar. If I tried to keep adding tick marks to the horizontal axis at $5,000 increments, as I did on the rest of the scale, it would extend the chart for dozens of pages.

Therefore, talking *just* about the average or median family saving $1,182 is nearly meaningless. Most households and families would save either less money or much more than that.

As a taxpayer myself, and as someone who enjoys having civil discussions, I worry about high taxes, but I also worry about balanced budgets and investment in infrastructure, defense, education, and health care. I care about both liberty and fairness. Therefore, I want to know from my representatives how much families all over the income spectrum would save thanks to the tax cut. In a case like this, we must do more than merely refer to a grossly simplified median or average; we must show more data.

How much, on average, will people with incomes of $10,000, $100,000, and $1,000,000 save every year?

The Tax Policy Center made the following estimate of the percentage increase in after-tax income that a typical household in each of several income brackets would enjoy thanks to the Tax Cuts and Jobs Act:[7]

I believe that it's worth debating whether it's fair that households mak-

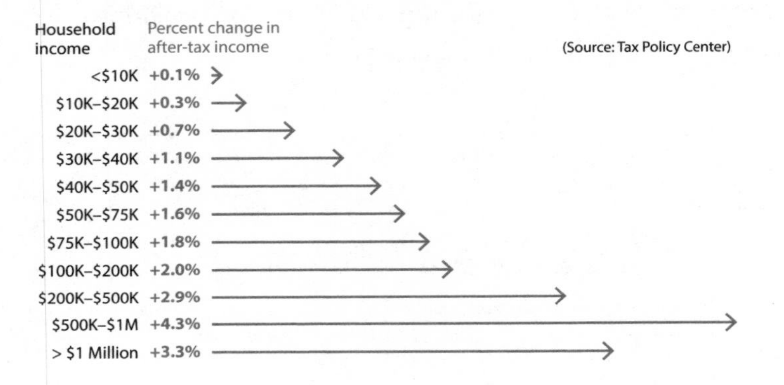

ing more than $1 million enjoy an increase of after-tax income of 3.3% (3.3% of $1 million is $33,000) while a middle-class family making, say, $70,000 gets an additional 1.6% ($1,120 per year). You're free to support or oppose the tax cut, but to deliberate about it, we must see data that is much more detailed than a mere average or median.[8] These measures of central tendency can be useful, but very often they don't summarize the shape and texture of a data set well. Charts based on averages alone lie sometimes because those averages capture too little information.

When discussing topics such as income, it is also possible to lie by showing *too much* information. Imagine that I could plot the income of every single household in the United States, ending up with a chart with tens of millions of little dots. That would, of course, be overkill. We don't need that level of detail to enable a conversation. The histogram of household income distribution provides a good balance between too much simplification and too much complexity, which is what we should demand from all charts we consume.

I love adventure movies, and Marvel's *Black Panther*, directed by Ryan Coogler, is a terrific adventure with a compelling plot and charismatic characters. It was also very successful at the box office, to the point that, according to many news outlets, it became "the third highest-grossing movie of all time in the U.S., behind *Star Wars: The Force Awakens* and *Avatar*."[9]

This isn't true. *Black Panther* was a deserving, smashing success, but it is very likely not the third highest-grossing movie ever in the United States.[10]

A common problem in stories about movie box offices is that they often contain nonadjusted prices when they should consider adjusted prices. I bet that you're paying more for goods today than you were five years ago. If you've stayed in the same job for many years, your salary may have increased, too. Mine has, but in *absolute* terms (nominal value), not in *relative* terms (real value). Because of inflation, even if my salary may look bigger when it enters my bank account every month, it may not *feel* bigger: the amount of stuff that I can purchase with it is roughly the same as it was three or four years ago.

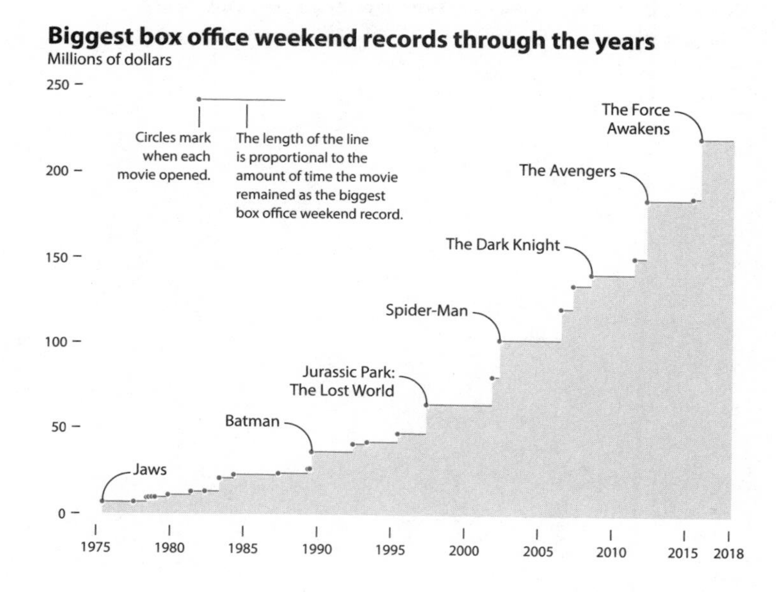

That's the challenge behind charts such as the one on the previous page, inspired by data analyst and designer Rody Zakovich,[11] who used data from the website Fandango. The chart plots the biggest box office opening weekends (note: Rody is well aware of the chart's shortcomings).

This chart—which shows first weekend box office records, not total over time, therefore *Black Panther* isn't in it—lies in the same way that most of the stories you see in social media touting the latest record-breaking movie lie: they often aren't adjusted for inflation, so they show nominal values, not real ones. It's easier to become "the highest-grossing movie" of all time if movie tickets cost $15 instead of $5, in unadjusted dollars. That's why in many rankings of movie box office, very recent movies are usually on top and older ones are at the bottom.

To correct for this, I converted each movie's box office in the chart into 2018 dollars using a free online tool created by the Bureau of Labor Statistics.[12] I then plotted the results, which look a bit different from the above chart. The ranking of first-week box offices doesn't change much, but older movies look much better. See for yourself:

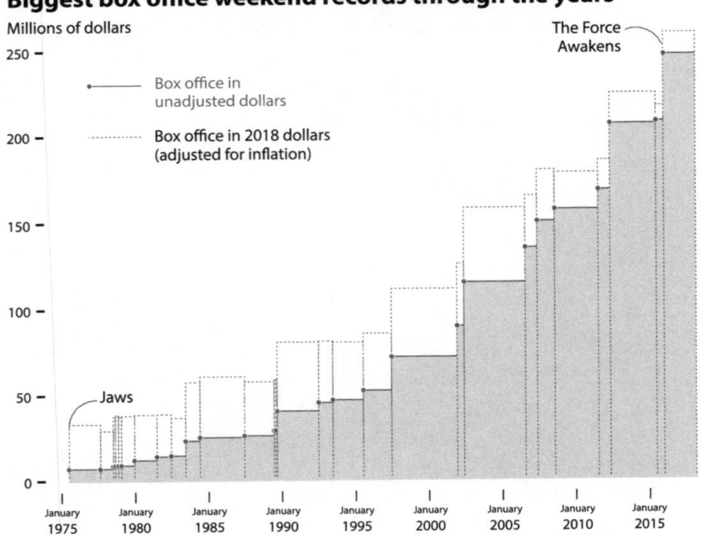

In the chart, I compare the unadjusted figures (red line) with the ones adjusted to 2018 dollars. All bars increase in height, but at very different rates: for *Star Wars: The Force Awakens* (2015), the box office changes by roughly 5%, whereas for *Jaws* (1975) it grows by more than 360%, which means that if it were released in 2018 it would make not its nominal box office of $7 million but $32 million.

I'm no expert in the economics of film production, just someone who enjoys movies and news, but as a professor and designer who studies charts for a living, I do find that charts and stories discussing successes and flops at the box office tend to be lacking. Isn't comparing *Jaws* with *The Force Awakens* unfair if we don't consider how much the movie industry has changed? And what about factors such as marketing and promotion efforts, the number of theaters in which each movie opened, and so forth?

I can't answer those questions, but I can use publicly available data to calculate how much each of the movies above made per theater during its opening weekend, and then I can transform the results into 2018 dollars:

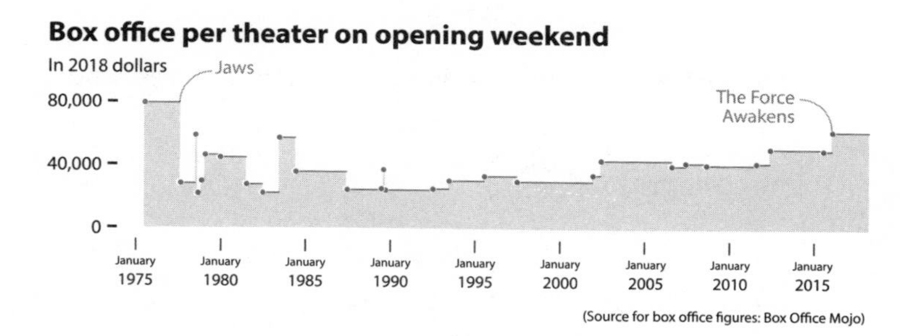

This made me wonder what would have happened if *Jaws*, which in 1975 opened in 409 theaters in the United States, had instead been released in 2015 in as many theaters as *The Force Awakens* (4,134)? Would releasing it in 10 times as many theaters have resulted in 10 times its nominal opening weekend box office, moving from $32 million to $320 million? Who knows? It could also be that modern movie theaters have smaller capacities, on average, than the ones in the 1970s. So many questions!

Other metrics of relative success would be profit (the difference between a movie budget and its total box office) and return on investment (the ratio between a movie's profit and its budget). Movies such as *Avatar, The Avengers,* and *The Force Awakens* were very profitable but at relatively high risk, as they cost a lot to produce and to promote. Some estimates say that nowadays you may need to spend as much on marketing as you do on making the movie itself. A 2012 movie designed to be a blockbuster, *John Carter,* cost Disney more than $300 million to make and market, yet it recovered just two-thirds of that.[13]

Other movies are much less risky: The movie with the highest return on investment of all time, according to some sources,[14] is *Paranormal Activity.* It made nearly $200 million but it cost only $15,000 to make (marketing costs excluded). Which movie is more successful, *Avatar* or *Paranormal Activity?* It depends on the metric we choose and on how we weigh returns versus possible risk on each investment.

Therefore, here's a new version of my chart. I calculated how much of each movie's budget—not counting marketing cost—was recovered during its first weekend:

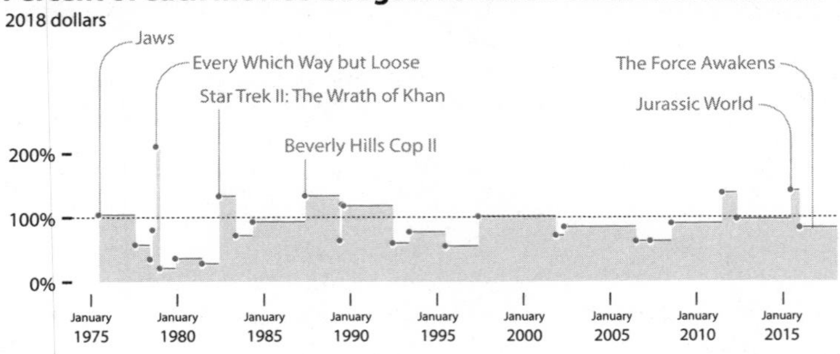

Jaws recovered its entire budget on the first weekend, while other movies made a profit at the outset. The most extreme case is a movie that made *twice* its entire budget at its opening: *Every Which Way but Loose*, in which Clint Eastwood partners up with an orangutan called Clyde. Now that I think of it, I loved that movie when I was a kid.

When it comes to designing a chart, which values are better, nominal (nonadjusted) or real (adjusted)? It depends. Sometimes, adjusted values matter much more. Comparing box offices or any other kind of price, cost, or salary over time doesn't make sense if you don't adjust your figures, as we've just seen. To understand a nominator, you need to pay attention to the denominator, particularly if you make comparisons between groups that have different denominators.

Imagine that I give you two slices from one pizza and I give another person three slices from a different pizza. Am I being mean to you? It depends on how many slices each of the pizzas is divided into:

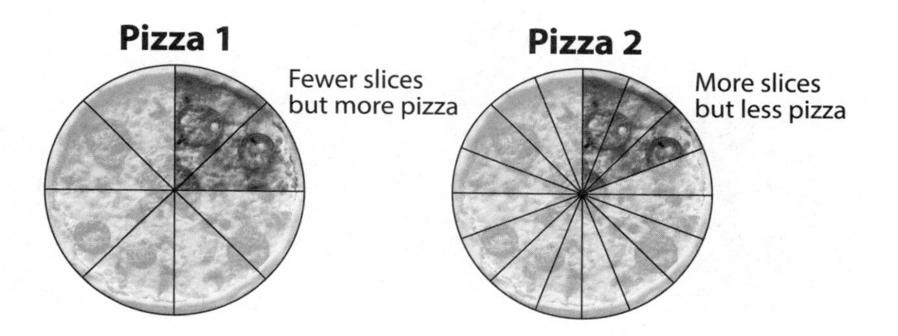

Not taking denominators into account can have grave consequences. Here's a bar graph inspired by fictional data provided by Judea Pearl in his *The Book of Why: The New Science of Cause and Effect*:

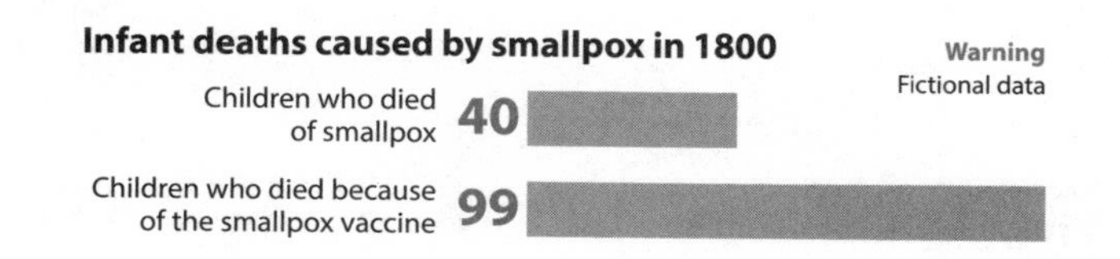

Pearl's fictional data reflects numbers thrown around during heated debates in the 19th century, when the smallpox vaccine became widespread, between those in favor of universal inoculation and those opposed. The latter were worried because the vaccine caused reactions in some children and those reactions sometimes led to deaths.

As alarming as it looks ("More children died because of the vaccine!"), the chart I designed isn't enough to help you make a decision about whether to vaccinate your own children. For it to tell the truth, I need it to display much more data, including the denominators. This flow and bubble chart can make us smarter at reasoning about this case:

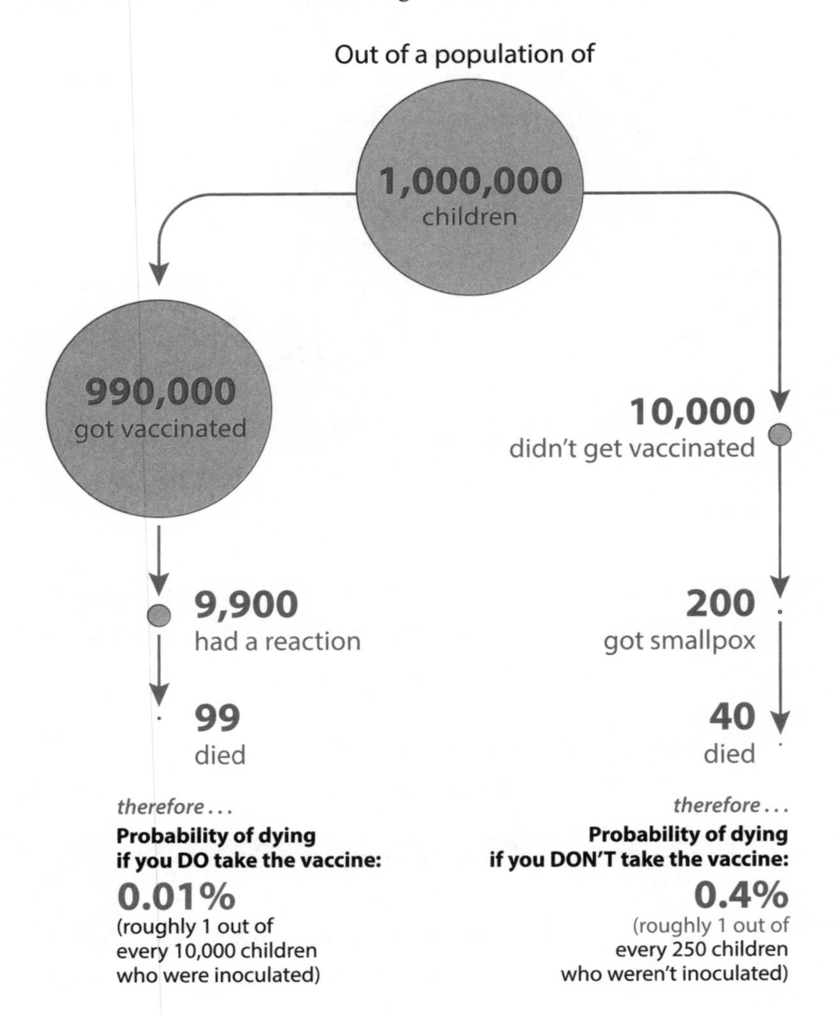

Let's verbalize what the chart reveals: 99% of children out of my fictional population of 1 million took the vaccine. The probability of having a reaction is roughly 1% (that's 9,900 out of 1 million). The probability of dying if you have a reaction is also 1% (99 out of 9,900). But the probability of dying because of the vaccine is just 0.01% (99 out of 990,000 who were inoculated).

On the other hand, if you don't take the vaccine, there's a 2% probability of getting smallpox (200 out of 10,000). And if you do get the disease, there's a 20% chance that you'll die (40 out of 200). The reason why my first chart shows that many more children died because of a reaction to the vaccine than because of smallpox itself is simply that the population that took the vaccine (990,000) is enormously larger than the population of children who didn't (10,000), a fact that I should have disclosed.

I agree that 99 versus 40 still looks like a huge difference, but try to reason with a hypothetical. Imagine that *no children* were inoculated against smallpox. We know that 2% will get the disease. That's 20,000 children out of a population of 1 million. Of those, 20% will die: 4,000 in total. Here's my updated chart:

Infant deaths caused by smallpox in 1800

Warning
Fictional data

Children who died because of smallpox or the vaccine against it **139**

Children who *would* die if the vaccine hadn't been widely applied **4,000**

The 139 there is the result of adding the 40 kids who weren't vaccinated and died of smallpox and the 99 who were vaccinated and died after a reaction to the vaccine. The comparison between universal inoculation and no inoculation at all is now more truthfully revealed.

In many cases, *both* the nominal *and* the adjusted values matter for different reasons. 100 People (https://www.100people.org/) is a wonderful website that translates many public health indicators into percentages. Out of every 100 people in the world, 25% are children, 22% are overweight, and 60% are of Asian origin. Here's a statistic that made me feel optimistic:

If the world were
100 people

1 would be dying of starvation

(Weepeople font, copyright © 2018 ProPublica and Alberto Cairo)

Data analyst Athan Mavrantonis pointed out that these numbers could be interpreted in a different way:

Which chart is better? The answer is neither. Both are relevant. It's true that the proportion of people suffering from starvation is small in relative terms—and it keeps shrinking—but it's also true that behind that 1% figure, there are *74 million human beings*. This is a bit less than the population of Turkey or Germany and roughly equivalent to one-quarter of the population of the United States. Now the chart doesn't look so bright, does it?

Several recent books cast a positive light on human progress. Hans Rosling's *Factfulness* and Steven Pinker's *The Better Angels of Our Nature* and *Enlightenment Now* contain an impressive array of statistics and charts corroborating that the world is indeed becoming a better place.[15] These books, and the websites they extract their data from, such as Our World in Data (https://ourworldindata.org), suggest that we may soon fulfill the United Nations' 2015 Global Goals initiative, which aims to "end poverty, fight inequality and stop climate change" by 2030. I do believe that charts like these, based on data from the World Bank, are fabulous news:

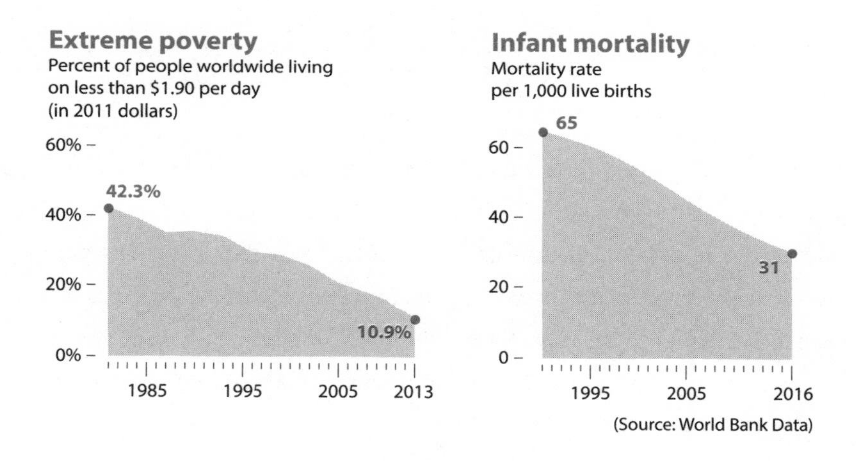

In 1981, around 4 out of 10 people in the world had to survive on the equivalent of less than two dollars per day. In 2013, that number dropped to 1 out of 10 people. In 1990, 65 children out of 1,000 died before reaching one year of age. In 2017, that number had dropped to 31.

This is a success story that must be celebrated. Whatever it is that institutions such as the United Nations, UNICEF, and many others are doing in collaboration with governments and nongovernmental organizations, it seems to be working, and it needs to continue.

However, charts and data like these may obscure the massive amount of human misery that hides behind the figures. Percentages and rates numb our empathy. A figure like 10.9% sounds small until you realize how many people it represents—close to 800 million in 2013:

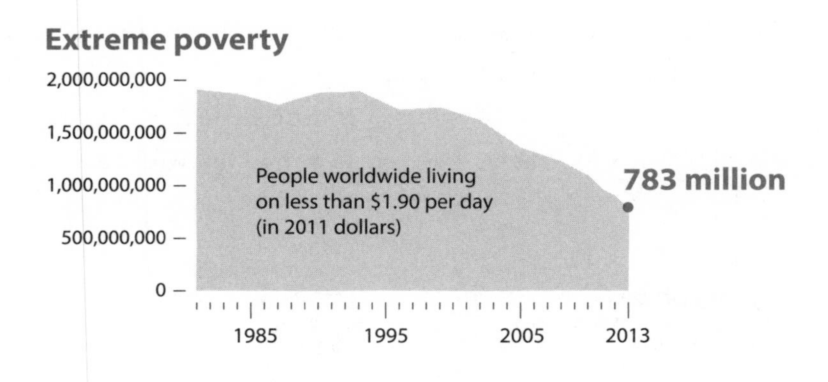

I think that seeing percentages and rates alone ("10.9% of the world population") when discussing human progress may make us too complacent, as it dehumanizes the statistics. I'm not alone in this hunch. Psychologist Gerd Gigerenzer, author of *Calculated Risks*, says that percentages make numbers more abstract than they should be. I'd suggest peeking into the raw counts, too, and reminding ourselves that "that's 783 million people!"

Neither the adjusted figures nor the nominal ones are enough if presented on their own. When shown together, they give us a richer understanding of the breathtaking progress we've made—and of the breathtaking challenges we still face. Nearly 800 million human beings in extreme poverty is the equivalent of two and a half times the population of the United States in 2016. That's a lot of suffering.

Many charts conceal relevant baselines or counterfactuals that, if revealed, would reverse an intended message. Take this 2017 tweet from WikiLeaks founder Julian Assange, accusing modernity of making advanced nations childless and more reliant on immigration:

> Capitalism+atheism+feminism = sterility = migration.
> EU birthrate = 1.6. Replacement = 2.1. Merkel, May, Macron, Gentiloni all childless.[16]

The leaders Assange mentioned are Angela Merkel, chancellor of Germany; Theresa May, the British prime minister; Emmanuel Macron, president of France; and Paolo Gentiloni, Italy's prime minister.

Assange illustrated his tweet with a table of data from more than 30 European countries. These are Assange's figures, in a chart that is as cluttered as the table itself:

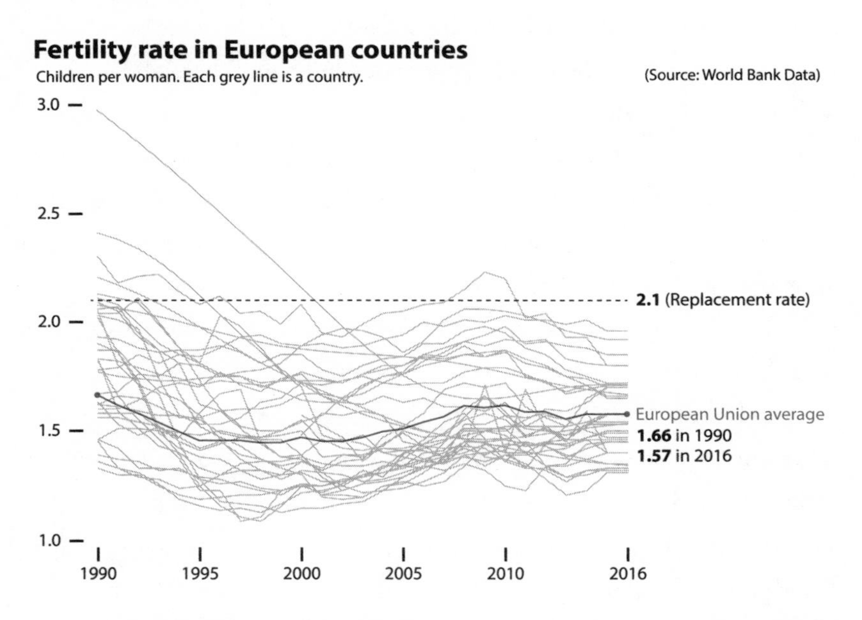

Assange made several mistakes. First, he wrote "birth rate," but the figures he used are *fertility* rates; the variables are somewhat related, but they aren't the same. The birth rate is the number of live births per 1,000 people in a country and year. The fertility rate is, to simplify a bit, the average number of children a woman could give birth to during her lifetime. If half the women in a country have 2 children, and the other half have 3, then the fertility rate of that country would be 2.5.

But let's ignore this mistake and assume that Assange wanted to write "fertility rate." With his tweet and data display he argued that the fertility rate in these capitalist and secular democracies is quite low—1.6 children per woman on average—and that the leaders of such countries may have something to do with the fact that this figure is far from the minimum fertility rate needed to keep a population stable in the long term, 2.1 children per woman—also called the replacement rate.

Assange's table and my chart version are spectacular because they accomplish two opposite feats: they lie by showing too little data *and* too much data—or at least, too much data that hinders rather than enables understanding.

Let's begin with the latter. Tables with many numbers and charts in which all the lines overlap so much, like mine, make it very hard to extract patterns from the data or to focus on specific cases that may refute our initial opinion. For instance, northwestern European countries tend to be quite secular and favor gender equality. Have their fertility rates dropped dramatically?

We could separate the lines, rather than cram them all into a single chart, and see what happens, on the facing page.

See Denmark or Finland. Their lines display very little variation since 1990, and they stay quite close to the replacement rate of 2.1. Now focus on more religious countries, such as Poland and Albania: the drops in their fertility rates are quite pronounced. Next, notice countries where a majority of the population claims to be Christian, such as Spain and Portugal. Their fertility rates are very far from the replacement rate.

This leads me to guess that the main factor driving changes in fertility rates in countries that have not experienced recent wars or catastrophes might not be religion or feminism, Assange suggested, but perhaps the stability of their economies and societal structures. For instance, unemploy-

Fertility rate in European countries, 1990–2016

Compared to the replacement rate: 2.1 children per woman

Note: not all these countries belong to the European Union.

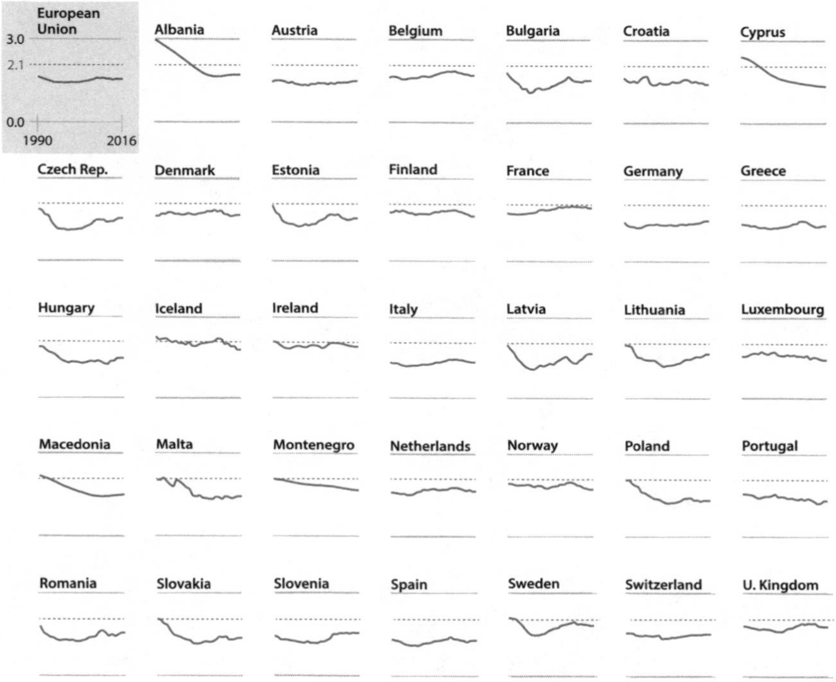

European Union
3.0
2.1
0.0
1990
2016
Albania
Austria
Belgium
Bulgaria
Croatia
Cyprus
Czech Rep.
Denmark
Estonia
Finland
France
Germany
Greece
Hungary
Iceland
Ireland
Italy
Latvia
Lithuania
Luxembourg
Macedonia
Malta
Montenegro
Netherlands
Norway
Poland
Portugal
Romania
Slovakia
Slovenia
Spain
Sweden
Switzerland
U. Kingdom

ment in southwestern European countries such as Spain, Italy, or Portugal has been historically high, and salaries are low; people may delay or abandon the decision to have children simply because they know they can't afford them. The large drops in some ex-Soviet countries such as Albania, Hungary, Latvia, or Poland in the early 1990s could be related to the fall of the Soviet Union in 1991 and the transition to capitalism.

Migration, as Assange seems to point out in his tweet, might help to increase fertility rates or make a country age more slowly, but to reach that conclusion we need more evidence. Assange's table and my corresponding charts fail by not showing enough data and context. We are cherry-picking. Fertility rates have dropped not only in secular nations but nearly *everywhere* in the world, in religious and secular countries alike.

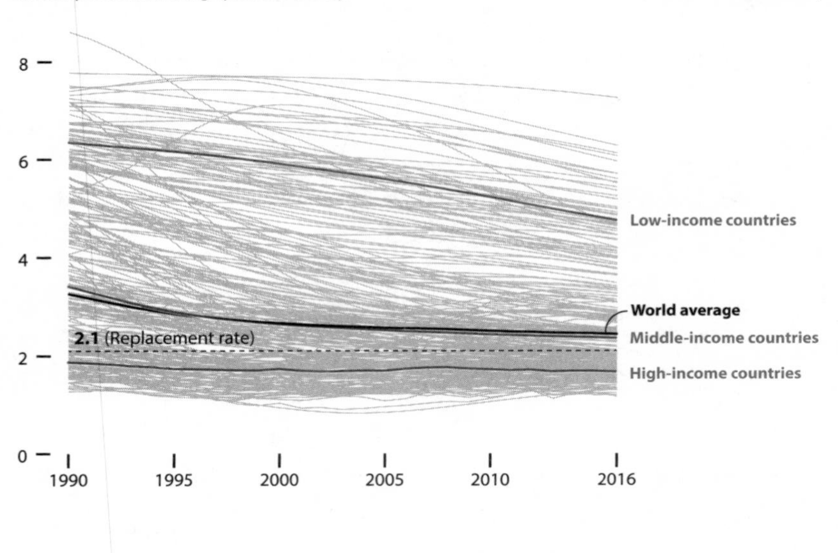

Let's end this chapter by circling back to the discussion about nominal values and unadjusted data versus rates and percentages. Did you know that the most obese places in the United States are Los Angeles County (California), Cook County (Illinois), and Harris County (Texas)?

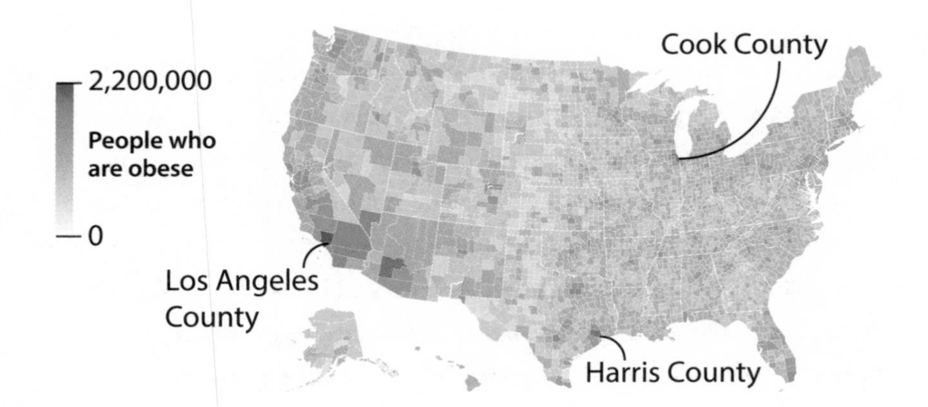

Coincidentally, those places are also the poorest in the nation:

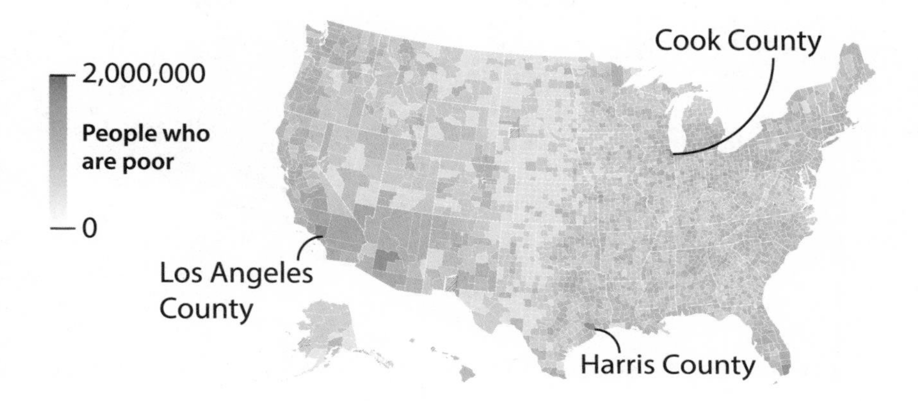

The relationship is remarkable—except that it truly isn't. Here's a map of county population:

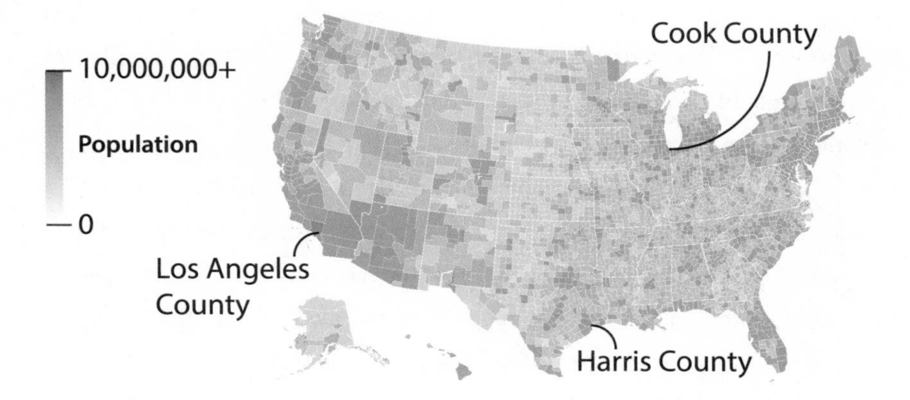

The number of obese people is strongly correlated to the number of poor people simply because both variables are also strongly correlated to the size of the underlying populations: Cook County is where the city of Chicago is, and Harris County contains Houston. Here are two maps with the figures transformed into percentages:

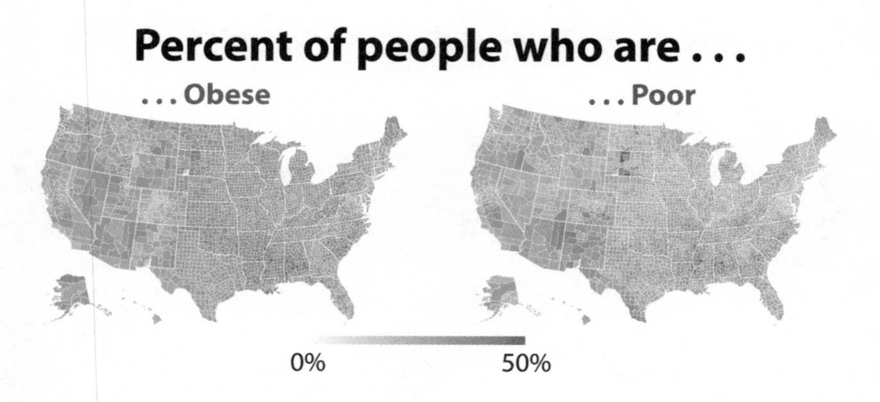

The picture changes quite a lot, doesn't it? There's still a vague relationship between obesity and poverty but it's much weaker, and counties such as Los Angeles are certainly not among the top counties. Los Angeles has so many poor and obese people because it has a big population. Maps that use shades of color to represent data—"choropleth" maps, from the Greek words *khōra* (place) and *plēthos* (crowd or multitude)—work better when they display adjusted data, such as percentages of obese and poor people, not raw counts. When they show raw counts they simply reflect the size of the population in the regions they depict.

We can visualize the numbers in a different way, as scatter plots. The first chart on the opposite page shows the relationship between obesity and poverty without adjusting by population; the second one shows the relationship between the percentage of people who are obese and the percentage of people who are poor. (See the charts on the next page.)

Claiborne County, in Mississippi, has the highest percentage of obese people (48% of its 9,000 inhabitants) and Oglala Lakota County, in South Dakota, has the highest percentage of people living in poverty (52% of its 13,000 inhabitants). Los Angeles County, Cook County, and Harris County have obesity rates of between 21% and 27% and poverty rates of between 17% and 19%. They are in the lower-left quadrant of the second chart.

This would be one case where both unadjusted and adjusted figures matter; after all, there are nearly two million poor people in Los Angeles County. But if your goal is to compare counties, the adjusted figures are necessary.

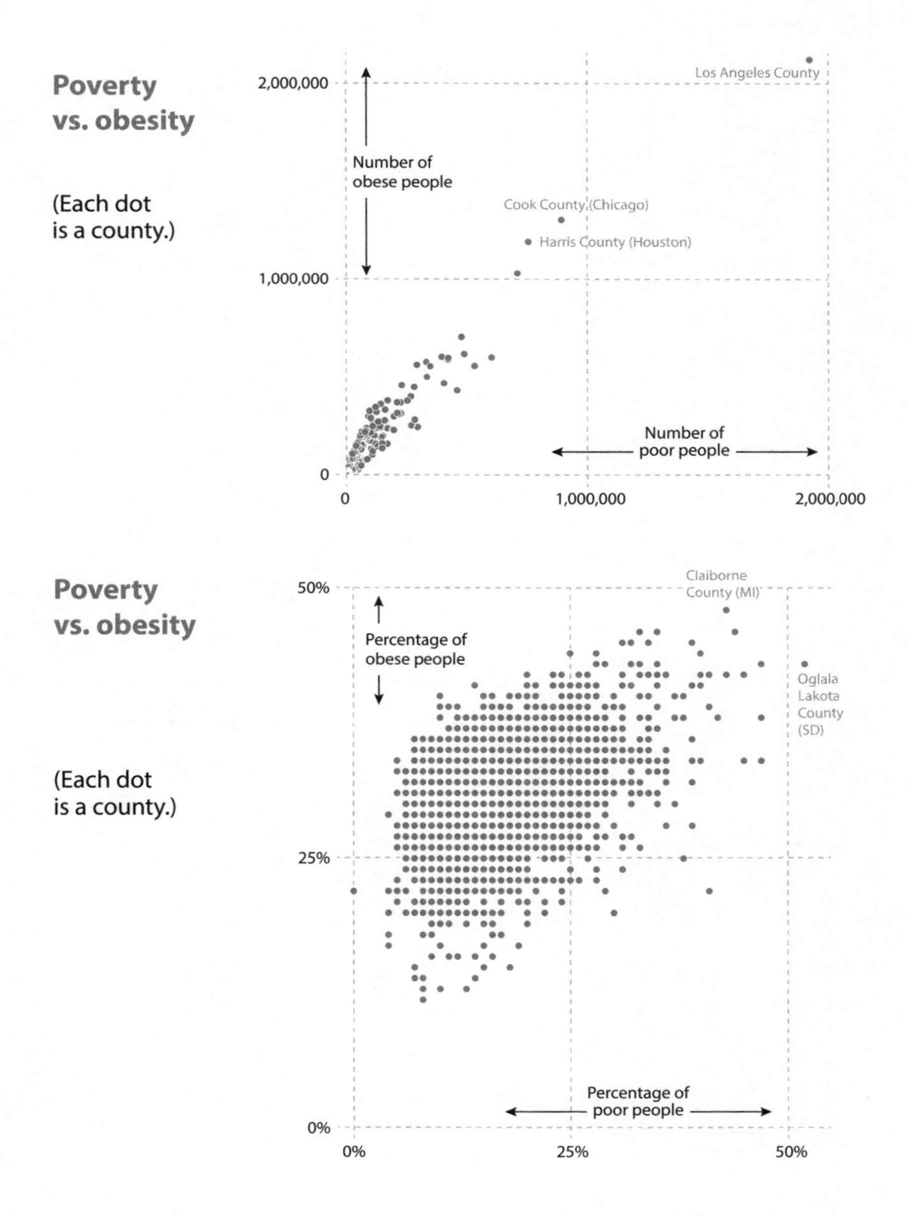

Chapter 5

Charts That Lie by Concealing or Confusing Uncertainty

To avoid lying, charts need to be precise; but sometimes too much precision is detrimental to understanding.

Data is often uncertain, and this uncertainty should be disclosed. Ignoring it may lead to faulty reasoning.

On the morning of April 28, 2017, I opened my copy of the *New York Times* to the Opinion pages to find Bret Stephens's first column. Stephens is a riveting conservative columnist who had been recruited from the *Wall Street Journal* to add some ideological variety to the *Times*'s commentary roster.

The title of Stephens's first piece was "Climate of Complete Certainty,"[1] and some of its lines were music to my ears: "We live in a world in which data convey authority. But authority has a way of descending to certitude, and certitude begets hubris." Unfortunately, other passages weren't that impressive. Later in the column, Stephens attacked the consensus on the basic science of climate change with bizarre arguments. For instance, he wrote (emphasis mine):

> Anyone who has read the 2014 report of the Intergovernmental Panel on Climate Change knows that, *while the modest (0.85 degrees Celsius, or about 1.5 degrees Fahrenheit) warming of the earth since 1880 is indisputable*, as is the human influence on that warming, much else that passes as accepted fact is really a matter of probabilities. That's especially true of the sophisticated but fallible models and simulations by which scientists attempt to peer into the climate future. To say this isn't to deny science. It's to acknowledge it honestly.

We'll get to the part about "fallible models and simulations" in a minute. For now, let's focus on the claim that an increase of 0.85 degrees Celsius worldwide is "modest." It sounds right. I don't think that anybody would feel any warmer if the temperature increased from 40°C to 40.85°C. The temperature would be perceived as equally hot.

However, if there's something that every citizen ought to know by now, it is that *weather* is not *climate*. That's why politicians who say that climate change isn't real because it's snowing heavily in their vicinity are either fooling you or ignorant of elementary school science. Stephens's "modest" increase of 0.85°C isn't modest at all if we put it into proper historical perspective. A good chart like the one on the opposite page can enable an informed conversation (BP stands for "before the present").[2]

Here's how to read this chart: on the horizontal axis are years, measured backward from the present year (year zero, on the right). The vertical axis is temperatures measured against the average temperature in degrees Celsius between the years 1961 and 1990, a common baseline in climate science; this becomes a horizontal dotted line on the chart. That's why we have positive temperatures (above that baseline) and negative temperatures (below that baseline).

The red line is the critical one; it's the result of averaging the tempera-

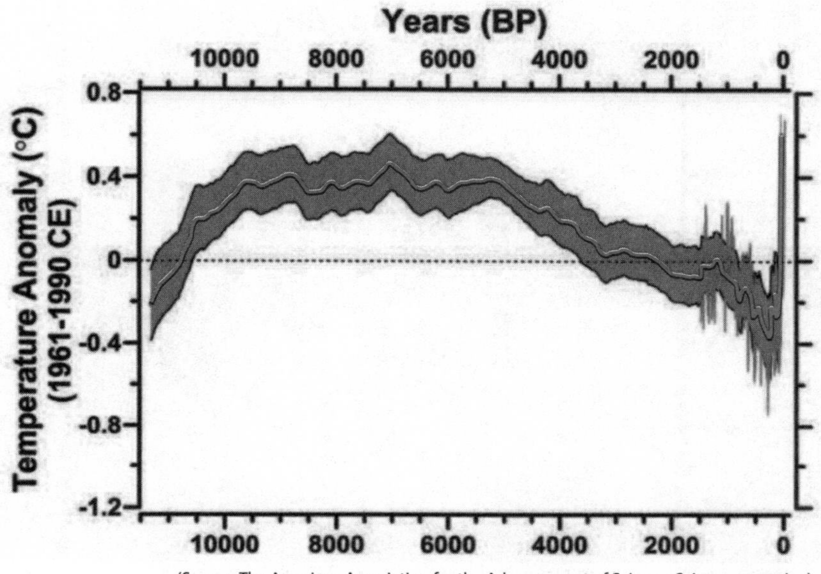

(Source: The American Association for the Advancement of Science, Science magazine)

ture variations of plenty of independent historical estimates done by competing researchers and teams from all over the world. The grey band behind it is the uncertainty that surrounds the estimate. Scientists are saying, "We're reasonably certain that the temperature in each of these years was somewhere within the boundaries of this grey band, and our best estimate is the red line."

The thin grey line behind the red one on the right side of the chart is a specific and quite famous estimate, commonly called the *hockey stick*, developed by Michael E. Mann, Raymond S. Bradley, and Malcolm K. Hughes.[3]

The chart reveals this: Contrary to what Stephens claimed, a 0.85°C warming isn't "modest." Read the vertical axis. In the past, it took *thousands of years* for an increase equivalent to the one we've witnessed in the past century alone. This becomes clearer if we zoom in and notice that no temperature change in the past 2,000 years has been this drastic:

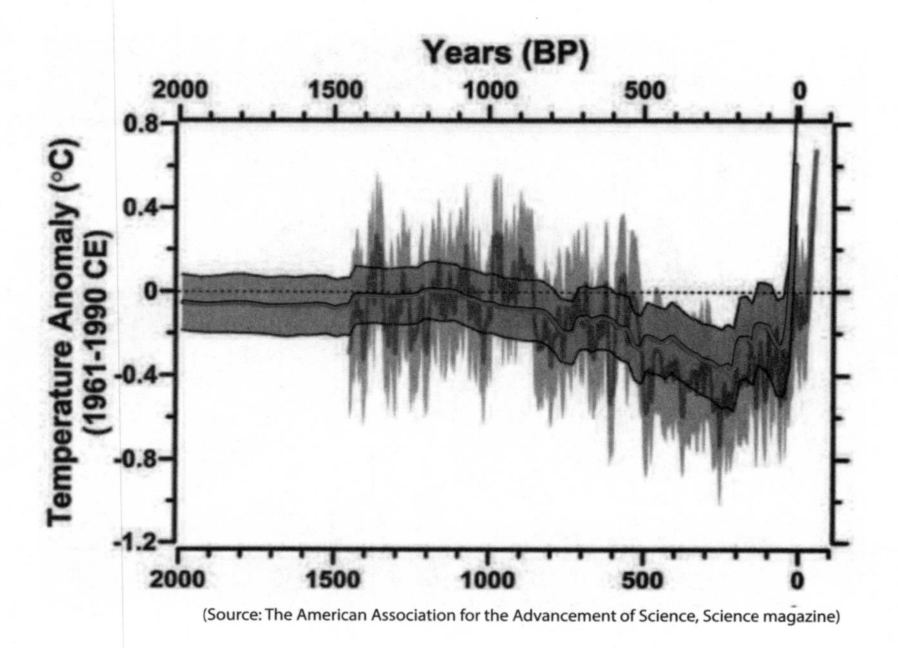

(Source: The American Association for the Advancement of Science, Science magazine)

There is no "modesty" after all.

What about Stephens's second claim, the one about the "fallible models and simulations"? He added:

> Claiming total certainty about the science traduces the spirit of science and creates openings for doubt whenever a climate claim proves wrong. Demanding abrupt and expensive changes in public policy raises fair questions about ideological intentions.

This all sounds like good advice at an abstract level, but not when we apply it to reality. First, climate models have been not only reasonably accurate but in many cases too optimistic.

The world is warming rapidly, ice sheets are melting, ocean and sea waters are expanding, and sea level is rising to the point that it may soon make life in regions like South Florida hard. Already, floods are becoming much more frequent in Miami Beach, even during good weather. This is leading the city to discuss those "expensive changes in public policy" that Stephens so much distrusts: installing huge water pumps and even elevating roads. This kind of conversation isn't based on "ideological" science—liberal or conservative—but on the facts, which can be observed directly.

See the following chart by the *Copenhagen Diagnosis* project.[4] It compares the forecasts made by the Intergovernmental Panel on Climate Change (IPCC) in the past to the actual recorded increase in sea levels:

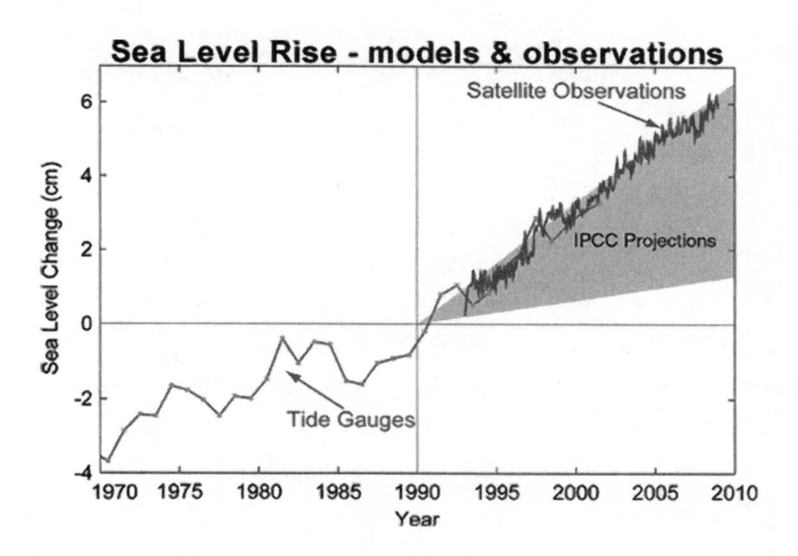

The grey band is the range of forecasts by the IPCC. Scientists were suggesting back in 1990 that sea level could rise roughly between 1.5 and 6.5 cm by 2010. Satellite observations—not "fallible models and simulations"—confirmed that the most pessimistic projection had turned out to be right. Have climate models been wrong in the past? Absolutely; science isn't dogma. Nonetheless, many of them have been grimly correct.

Finally, a critical point that Stephens failed to mention in his opening column at the *New York Times* is that even if data, models, forecasts, and simulations are extremely uncertain—something that climate scientists *always* disclose in their charts—they still all point in the same direction, with no exception. Here's one final good chart from the IPCC that Stephens could have showcased:

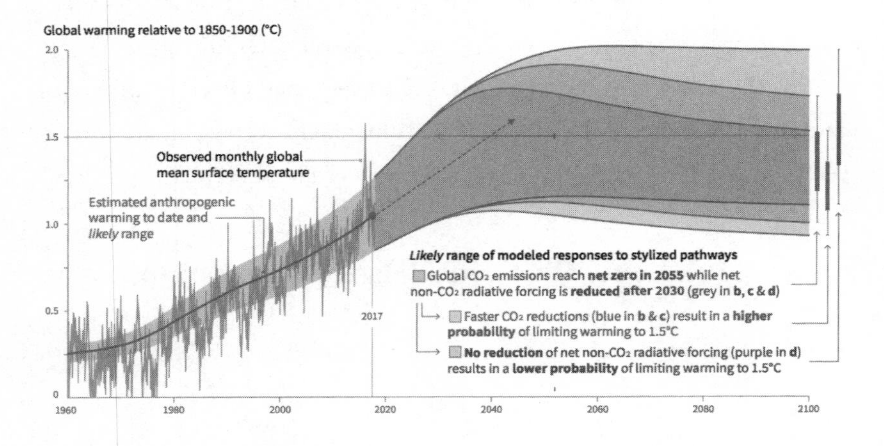

The chart shows several forecasts with their corresponding ranges of uncertainty. If you want to be extremely optimistic, the best available evidence suggests that global temperatures may increase as little as 1°C by 2100. That is a big increase; but worse, it's also possible that they will increase 2°C or more. There's still a possibility that we'll witness no further warming in the future, but it's equally likely that global temperatures will increase much more than 2°C, which would make the diminished parts of the Earth not covered in water much less habitable, as they'd be punished by extreme weather events, from monstrous hurricanes to devastating droughts.

An analogy would be useful here: if these weren't climate forecasts but probabilities of your developing cancer in the future calculated by several independent groups of oncologists from all over the world, I'm certain you'd try to take precautions right away, not ignore the imperfect—and only—evidence because it's based on "fallible models." All models are fallible, incomplete, and uncertain, but when all of them tell a similar story, albeit with variations, your confidence in them ought to increase.

I'm all for discussing whether the expensive changes in public policy mentioned by Stephens are worth it, but to have that conversation in the first place, we must read the charts and understand the future they imply. Good charts can help us make smarter decisions.

Bret Stephens's column is a good reminder that whenever we deal with data, we must take into consideration how uncertain estimates and forecasts are and then decide whether that uncertainty should modify our perceptions. We're all used to seeing polls reported like this:

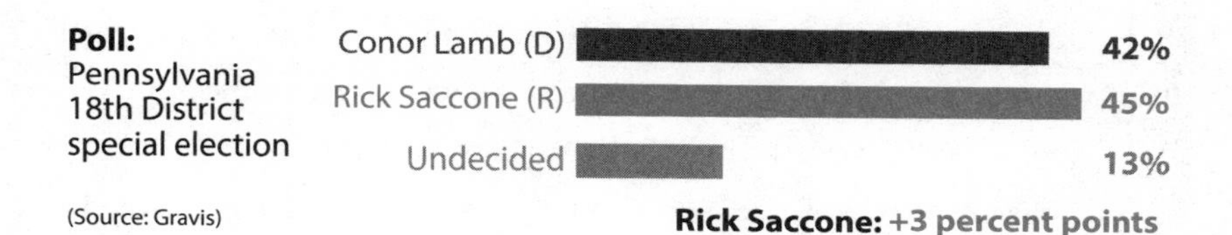

And then we are surprised—or upset, depending on the candidate we favored—when the final results look like this:

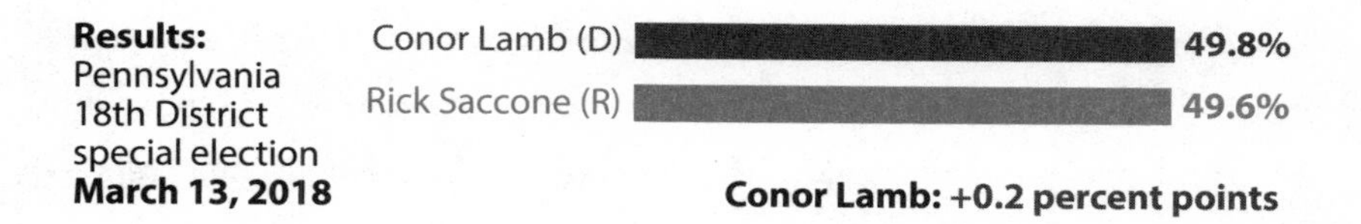

This comparison between a poll and the final results of a special election is useful for explaining that there are two different kinds of uncertainty lurking behind any chart: one can be easily computed, and another is hard to assess. Let's begin with the former. What these charts don't show—even if it's discussed in the description of the research—is that any estimate is always surrounded by error.

In statistics, "error" is not a synonym for "mistake," but rather a synonym for "*uncertainty*." Error means that any estimate we make, no matter how precise it looks in our chart or article—"This candidate will obtain 54% of the vote," "This medicine is effective for 76.4% of people 95% of the time," or "The probability of this event happening is 13.2%"—is usually a middle point of a range of possible values.

There are many kinds of error. One is the margin of error, common as a measure of uncertainty in polls. A margin of error is one of the two elements of a *confidence interval*. The other element is a *confidence level*, often 95% or 99%, although it can be any percentage. When you read that an estimate in a poll, scientific observation, or experiment is, say, 45 (45%, 45 people, 45 whatever) and that the margin of error reported is +/−3 at the 95% confidence level, you should imagine scientists and pollsters saying something that sounds more like proper English: Given that we've used methods that are as rigorous as possible, we have 95% confidence that the value we're trying to estimate is between 42 and 48, that is, 3 points larger or smaller than 45, which is our best estimate. We can't be certain that we have obtained a correct estimate, but we can be confident that if we ran the poll many times with the exact same rigorous methods we've used, 95% of the time our estimate would be within the margin of error of the truth.

Therefore, whenever you see a chart accompanied by some level of numerical uncertainty, you must force yourself to see a chart like the one below, always remembering that there's a chance that the final results will be even larger or smaller than the ones shown. The gradient areas represent the width of the confidence interval, which in this case is +/−3 points around the point estimates.

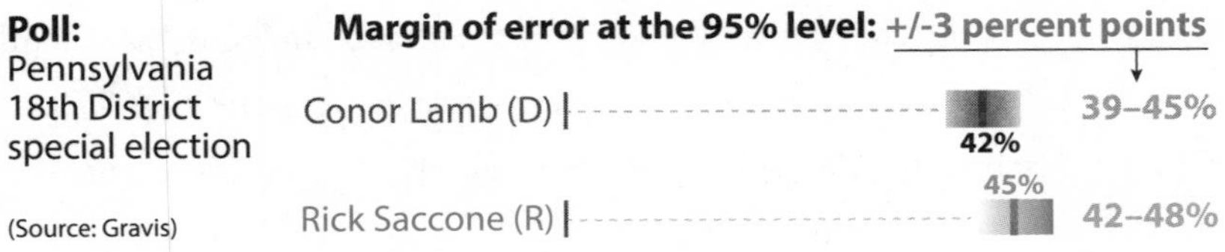

Poll:
Pennsylvania
18th District
special election

(Source: Gravis)

Margin of error at the 95% level: +/-3 percent points

Conor Lamb (D) 42% 39–45%

Rick Saccone (R) 45% 42–48%

Most traditional charts, such as the bar graph or the line graph, can be misleading because they look so accurate and precise, with the boundaries of the bars and lines that encode the data appearing crisp and sharp. But we can educate ourselves to overcome this design shortcoming by mentally blurring those boundaries, particularly when estimates are so close to each other that their uncertainty ranges overlap.

There is a second source of uncertainty on my first chart: the 13% of people who were undecided when the poll was conducted. This is a wild card, as it's hard to estimate which proportion of those would end up voting for either of the candidates. It can be done, but it would involve weighing factors that define the population you are polling, such as racial and ethnic makeup, income levels, past voting patterns, and others—which would yield more estimates with their own uncertainties! Other sources of uncertainty that are hard or even impossible to estimate are based on the soundness of the methods used to generate or gather data, possible biases researchers may introduce in their calculations, and other factors.

Uncertainty confuses many people because they have the unreasonable expectation that science and statistics will unearth precise truths, when all they can yield is imperfect estimates that can always be subject to changes and updates. (Scientific theories are often refuted; however, if a theory has been repeatedly corroborated, it is rare to see it debunked outright.) Countless times, I've heard friends and colleagues try to end a conversation saying something like, "The data is uncertain; we can't say whether any opinion is right or wrong."

This is going overboard, I think. The fact that all estimates are *uncertain* doesn't mean that all estimates are *wrong*. Remember that "error" does not necessarily mean "mistake." My friend Heather Krause, a statistician,[5] once told me that a simple rephrasing of the way specialists talk about the uncertainty of their data can change people's minds. She suggested that instead of writing, "Here's what I estimate, and here's the level of uncertainty surrounding that estimate," we could say, "I'm pretty confident that the reality I want to measure is captured by this point estimate, but reality may vary within this range."

We should be cautious when making proclamations based on a single poll or a specific scientific study, of course, but when several of them corroborate similar findings, we should feel more confident. I love reading about politics and elections, and one mantra that I repeat to myself is that any single poll is always noise, but the average of many polls may be meaningful.

I apply a similar principle when reading about unemployment, economic growth, or any other indicator. Usually, a single week's or month's variation may not be worth your attention, as it could be in part the product of reality's inherent randomness:

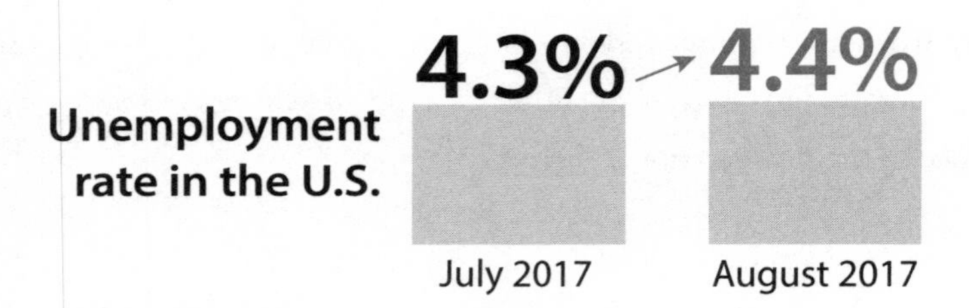

Zooming out enables you to see that the trend is exactly the opposite, since unemployment peaked in 2009 and 2010 and declined from there, with just some blips along the way. The overall trend has been one of steady decline:

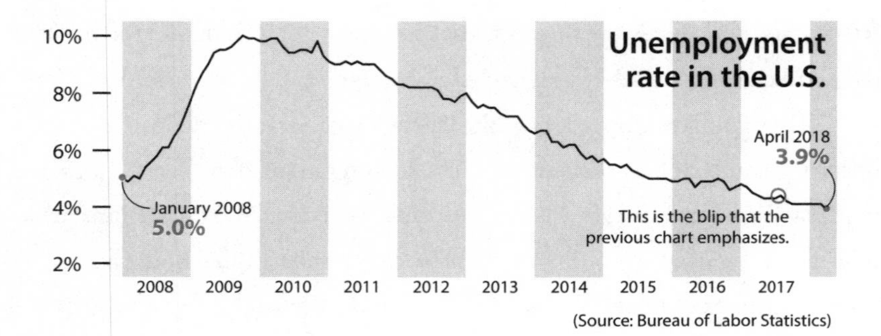

Even when confidence and uncertainty are displayed on a chart, they can still be misinterpreted.

I love coincidences. I'm writing this chapter on the very same day—May 25, 2018—that the National Hurricane Center (NHC) announced that subtropical storm Alberto was forming in the Atlantic Ocean and approaching the United States. Friends began bombarding me with jokes quoting NHC press releases: "Alberto meandering over the northwestern Caribbean Sea" and "Alberto is not very well organized this morning." Well, sure, I haven't gotten enough coffee yet today.

I visited the NHC's website to see some forecasts. Take a look at the chart below, plotting the possible path of my namesake storm. Down here in South Florida, we're very used to seeing this kind of map in newspapers, websites, and on TV during hurricane season, between June and November every year:

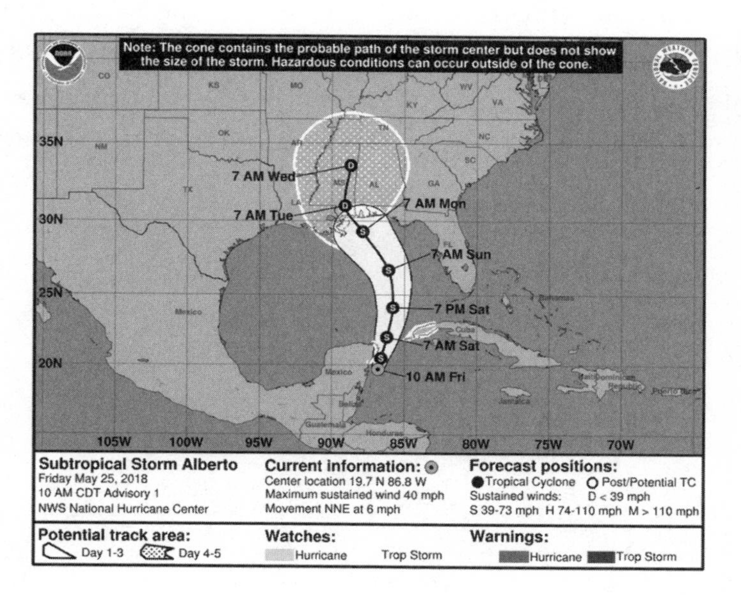

Years ago, my University of Miami friends Kenny Broad and Sharan Majumdar, experts on weather, climate, and environmental science, opened my eyes to how nearly everybody who sees this map reads it wrong; we're all now part of a multidisciplinary research group working to improve storm forecast charts, led by our colleague Professor Barbara Millet.[6]

The cone at the center of the map is popularly known as the *cone of uncertainty*. Some folks in South Florida favor the term "cone of death" instead, because they think that what the cone represents is the area that may be affected by the storm. They see the cone and they envision the scope of the storm itself, or the area of possible destruction, even if the caption at the top of the chart explicitly reads: "The cone contains the probable path of the storm center but does not show the size of the storm. Hazardous conditions can occur outside of the cone."

Some readers see the upper, dotted region of the cone and think it represents rain, although it simply shows where the center of the storm could be four to five days from now.

One good reason many people make these mistakes is that there's a pictorial resemblance between the cone, with its semicircular end point, and the shape of a storm. Hurricanes and tropical storms tend to be nearly circular because the strong winds make clouds swirl around their center. When seeing the cone of uncertainty, I always need to force myself *not* to see the following:

Journalists are also misled by the cone-of-uncertainty map. When Hurricane Irma was approaching Florida in September 2017, I remember hearing a TV presenter saying that Miami may be out of danger because the cone of uncertainty was on the west coast of Florida, and Miami is on the southeastern side, and therefore it was outside of the cone boundaries. This is a dangerous misreading of the map.

How do you read the cone of uncertainty correctly? It's more complicated than you think. The basic principle to remember is that the cone is a simplified representation of a range of possible paths the center of the storm could take, the best estimate being represented by the black line in the middle. When you see a cone of uncertainty, you should envision something like this instead (all these lines are made up):

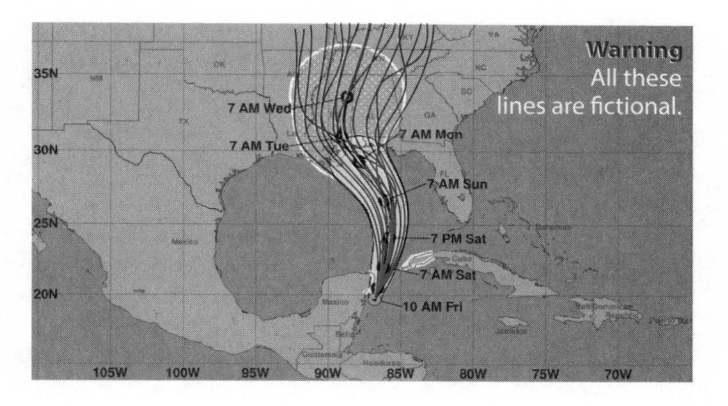

To design the cone, scientists at the NHC synthesize several mathematical models of where the current storm could go, represented by the imaginary lines in the first step (1) of the explainer below. Next, based on the confidence the forecasters have in the different models, they make their own prediction of the position of the center of the storm over the following five days (2).

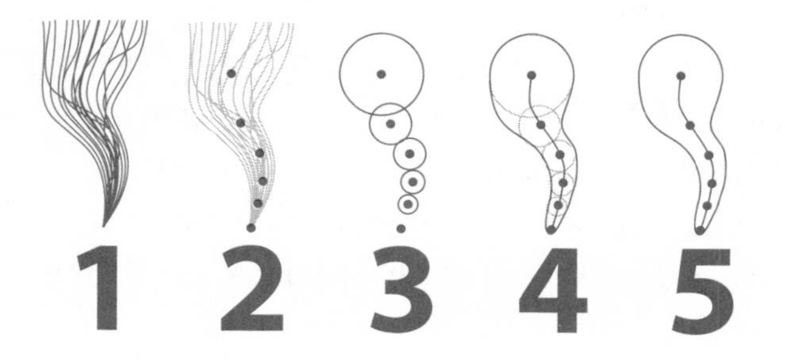

After that, they draw circles of increasing size (3) around the position estimates for each point. These circles represent NHC's uncertainty around those predicted positions. This uncertainty is the average error in all storm forecasts from the previous five years. Finally, scientists use computer software to trace a curve that connects the circles (4); the curve is the cone (5).

Even if we could visualize a map that looks like a dish of swirly spaghetti, it would tell us nothing about which areas could receive strong winds. To find that information, we would need to mentally overlay the sheer size of the storm itself, and we would end up with something that looks like cotton candy:

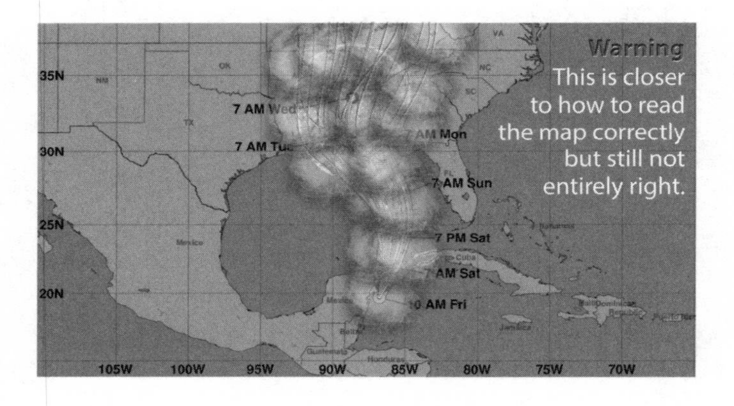

Moreover, we could wonder, "Will the cone always contain the actual path of the storm?" In other words, are forecasters telling me that out of the 100 times a storm like this moves through, under the same conditions of winds, ocean currents, and air pressure, the path of the storm will always lie within the boundaries of the cone of uncertainty?

Knowing a bit about numbers, that wasn't my assumption. Mine was something like this: 95 times out of 100, the path of the storm center will be inside the cone, and the line at the center is scientists' best estimate. But sometimes we may experience a freak storm where conditions vary so wildly that the center of the storm may end up running outside the cone:

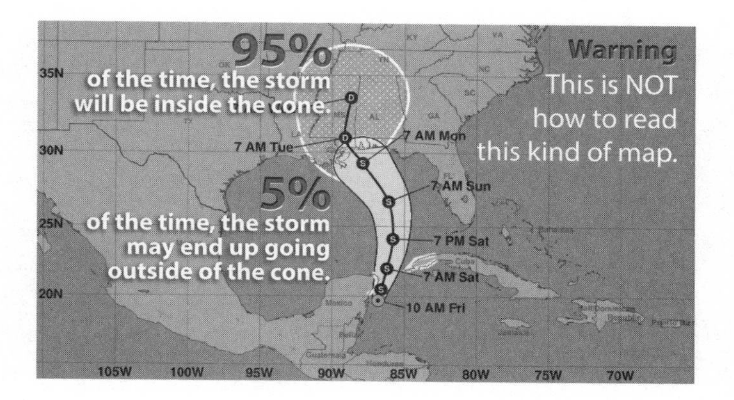

This is an assumption most people trained in science, data, and statistics would make. Unfortunately, they are wrong. According to rates of successes and failures at predicting the path of tropical storms and hurricanes, we know that the cone isn't designed to contain the central location 95% of the time—but just 67%! In other words, one out of three times that we experience a storm like my namesake, the path of its center could be outside the boundaries of the cone, on either side of it:

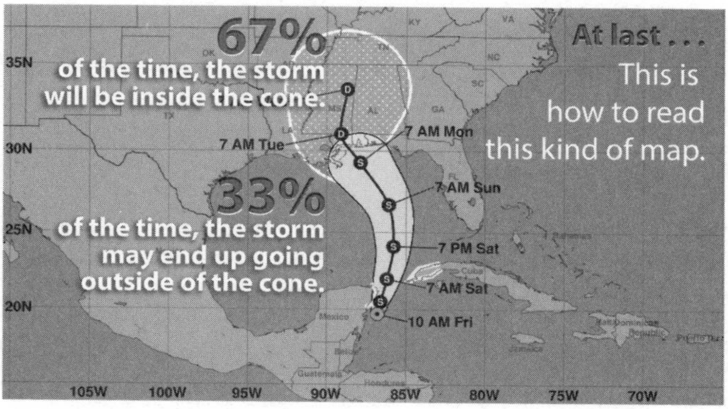

If we tried to design a map encompassing 95 out of 100 possible paths, the cone would be much wider, perhaps looking something like this:

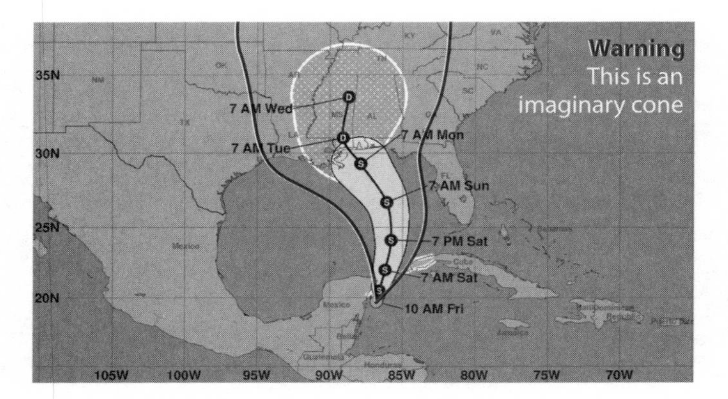

Adding the size of the storm on top of that, to offer a more accurate picture of which areas may be affected by it, will yield something that would surely provoke pushback from the public: "Hmm, this storm can go anywhere; scientists know nothing!"

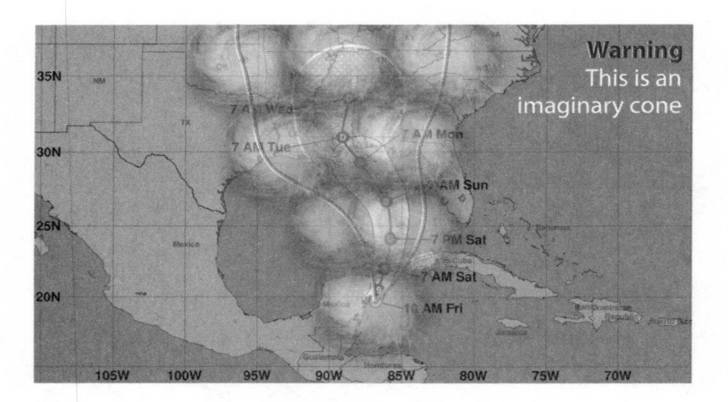

I warned against this kind of nihilism a few pages back. Scientists do know quite a bit, their predictions tend to be pretty accurate, and they get better every year. These forecast models, running on some of the world's largest supercomputers, are continuously improving. But they can't be perfect.

Forecasters always prefer to err on the side of caution rather than overconfidence. The cone of uncertainty, if you know how to read it correctly, may be useful for making decisions to protect yourself, your family, and your property, but only if you pair it with other charts that the National Hurricane Center produces. For instance, since 2017 the NHC releases this page with "key messages" about all storms:

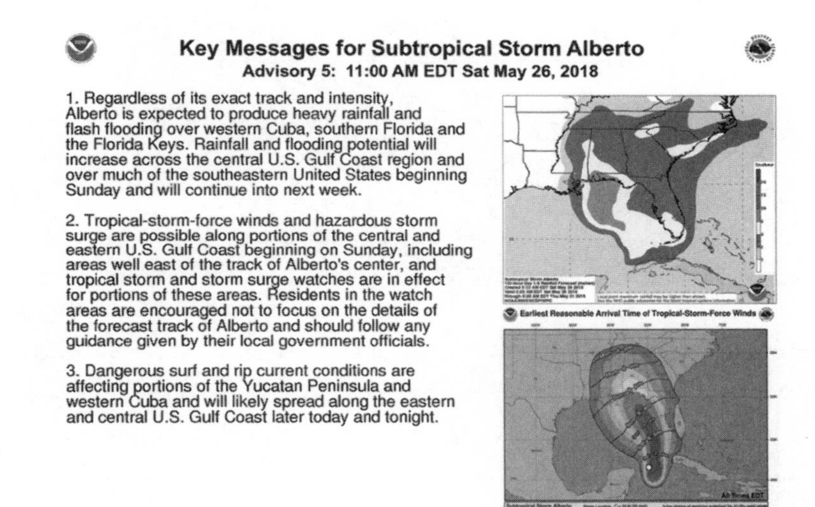
Key Messages for Subtropical Storm Alberto
Advisory 5: 11:00 AM EDT Sat May 26, 2018

1. Regardless of its exact track and intensity, Alberto is expected to produce heavy rainfall and flash flooding over western Cuba, southern Florida and the Florida Keys. Rainfall and flooding potential will increase across the central U.S. Gulf Coast region and over much of the southeastern United States beginning Sunday and will continue into next week.

2. Tropical-storm-force winds and hazardous storm surge are possible along portions of the central and eastern U.S. Gulf Coast beginning on Sunday, including areas well east of the track of Alberto's center, and tropical storm and storm surge watches are in effect for portions of these areas. Residents in the watch areas are encouraged not to focus on the details of the forecast track of Alberto and should follow any guidance given by their local government officials.

3. Dangerous surf and rip current conditions are affecting portions of the Yucatan Peninsula and western Cuba and will likely spread along the eastern and central U.S. Gulf Coast later today and tonight.

For more information go to hurricanes.gov

The page contains a map of probable rainfall (top) measured in inches, and another one of "earliest reasonable arrival time of tropical-storm-force winds" (bottom) that also displays the probability of experiencing them at all—the darker the color, the higher the probability.

Depending on the characteristics of each storm, the NHC includes different maps on the Key Messages page. For instance, if the storm is approaching the coast, the NHC may include maps showing the probability of experiencing storm surge and flooding. On the following page is a fictional map that the NHC provides as an example (and that is better seen in full color).[7]

These visuals aren't perfect. As you may notice, they don't reproduce well in black and white, and the color palettes and labeling they employ are sometimes a bit obscure. However, when seen side by side, they are much better

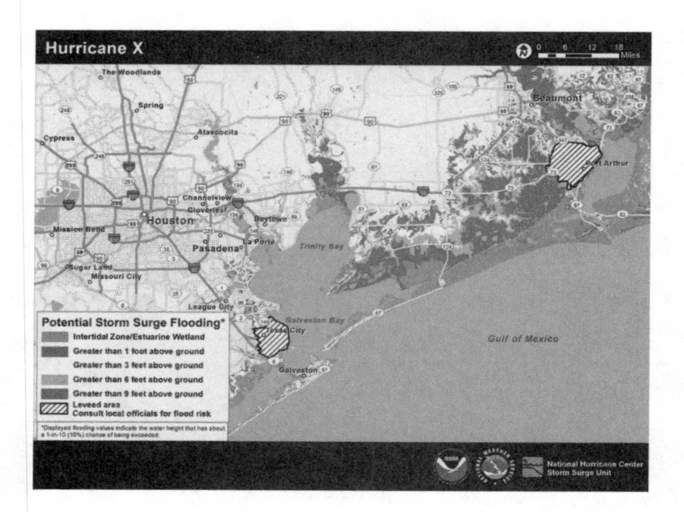

than the cone of uncertainty as tools to help us decide what to do when facing a storm.

These additional forecast charts are rarely showcased in the news, particularly on TV. I'm not sure of the reasons why, but my conjecture is that journalists favor the cone because it looks so simple, clear-cut, and easy to understand—albeit deceptively so.

The cone-of-uncertainty map lies to so many people not because it misrepresents uncertainty, but because it depicts data in a way that was not designed to be read by the general public. The map's intended audience is specialists—trained emergency managers and decision makers—even though any regular citizen can visit the NHC's website and see it, and news media uses it constantly. The cone is, I think, an illustration of a key principle: the success of any chart doesn't depend just on who designs it, but also on who *reads* it, on the audience's graphicacy, or graphical literacy. If we see a chart and we can't interpret the patterns revealed in it, that chart will mislead us. Let's now turn our attention to this challenge.

Chapter 6

Charts That Lie by Suggesting Misleading Patterns

Good charts are useful because they untangle the complexity of numbers, making them more concrete and tangible. However, charts can also lead us to spot patterns and trends that are dubious, spurious, or misleading, particularly when we pair them with the human brain's tendency to read too much into what we see and to always try to confirm what we already believe.

The eminent statistician John W. Tukey once wrote that "the greatest value of a picture is when it forces us to notice what we never expected to see."[1] Good charts reveal realities that may otherwise go unnoticed.

However, charts can also trick us into perceiving features that are meaningless or misleading. Did you know, for example, that the more cigarettes people consume, the more they live? This runs contrary to decades of evidence of the dangers of tobacco consumption—particularly cigarettes—but below you have a chart based on publicly available data from the World Health Organization and the United Nations.[2]

If I were a cigarette smoker, I'd likely feel reassured by this chart.

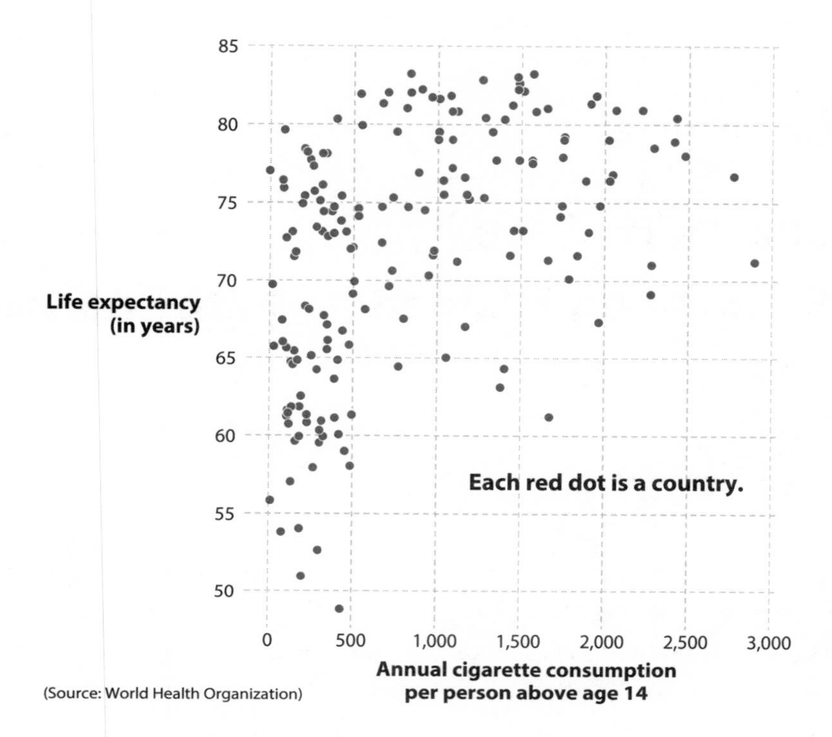

Tobacco doesn't reduce life expectancy! As unlikely as it sounds, the opposite might be true! My interpretation of this chart, though, is an example of several common challenges in chart reading: the relationship between correlation and causation, the role of amalgamation paradoxes, and the ecological fallacy. Let's examine them one by one.[3]

There's nothing wrong with the data plotted here, but my description of the chart ("The more cigarettes we consume, the more we live") is wrong. Describing the content of a chart correctly is critical. All this chart shows is that, at least at a country-by-country level, there's a positive association between cigarette consumption and life expectancy, and vice versa. But that doesn't mean that cigarettes increase life expectancy. Based on this example and others we'll soon see, we can enunciate a core rule of chart reading:

A chart shows only what it shows, *and nothing else.*

As I explained in chapter 1, "Correlation doesn't equal causation" is an old phrase repeated in elementary statistics courses. Correlation is usually one of the first clues to later finding causal links between phenomena, but the classic statistics saying has undeniable wisdom. The saying applies to this case, as there could be other factors that I'm not contemplating and that may influence both cigarette consumption and life expectancy. For instance, take wealth: People in wealthier countries tend to live longer because they usually have access to better diets and health care and are less likely to be the victims of violence and wars. Plus, they can afford to buy more cigarettes. Wealth could be a *confounding* factor in my chart.

The second and third challenges I mentioned before—amalgamation paradoxes and the ecological fallacy—are related. The ecological fallacy consists of trying to learn something about individuals based on characteristics of the groups they belong to. We saw it already when I mentioned that, despite being born in Spain, I'm far from being your average, stereotypical Spanish male.

The fact that people in one country smoke a lot and also live long doesn't mean that *you or I* can smoke a lot and also live long. Different levels of analysis—the individual versus the group—may require different data sets. If my data is created and summarized to study a group—countries, in this case—its utility will be very limited if what we need to study is either smaller groups, regions or cities of those countries, or individuals living in those places.

This is where amalgamation paradoxes come in. They are based on the fact that certain patterns or trends often disappear or even reverse depending on how we aggregate or subset the data.[3]

Considering that wealth may be a confounding factor in my first chart, let's draw it again with colors for different high-, middle-, and low-income groups:

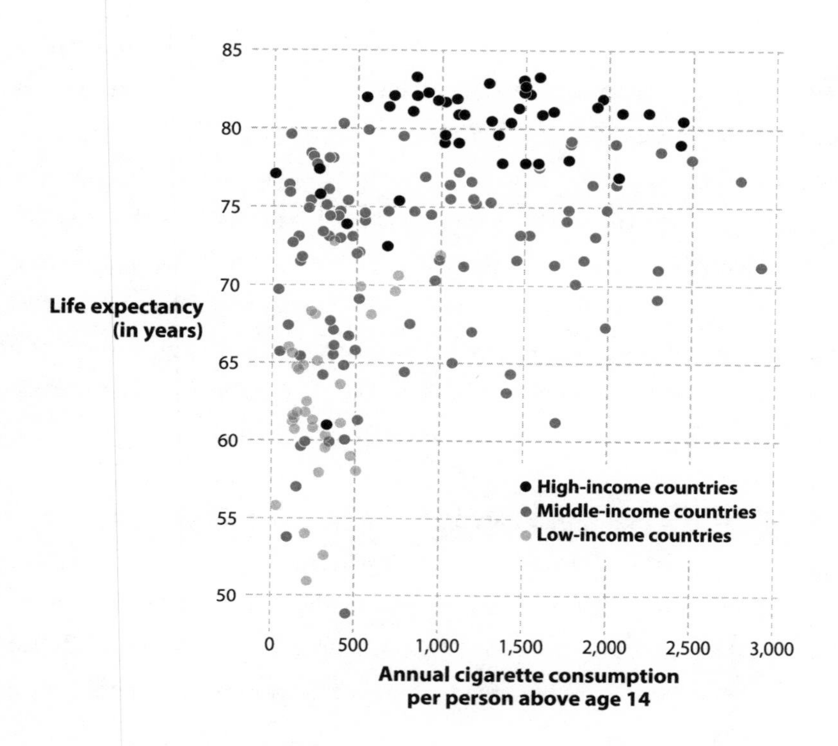

The chart looks quite messy, as countries of different income groups overlap too much, so let's split up the countries by income:

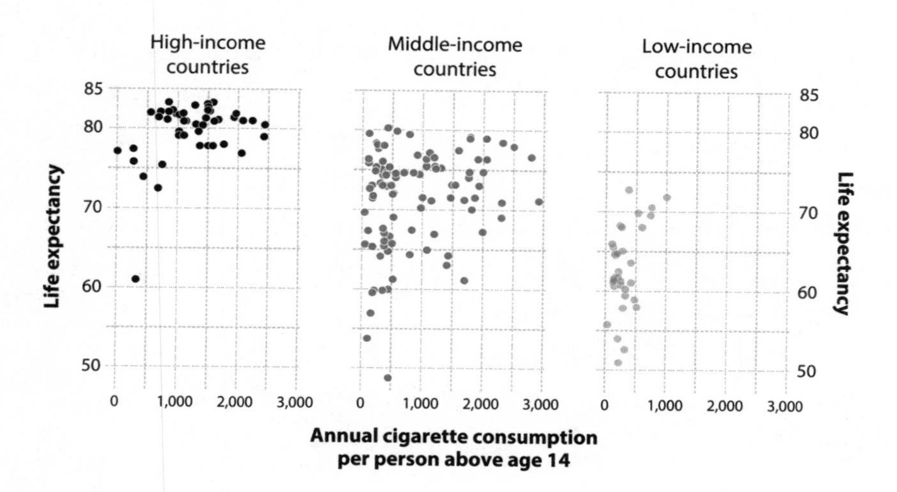

The strong positive association between cigarette consumption and life expectancy doesn't look that strong anymore, does it? Poor nations have a high variability in life expectancies (vertical axis) but, on average, they don't smoke that much. Middle-income countries have a large variation in both life expectancy and cigarette consumption, and the relationship between those variables is weak. High-income nations tend to have high life expectancies overall (they're high on the vertical axis), but cigarette consumption (horizontal axis) is all over the place: cigarette consumption is high in some countries and low in others.

The picture becomes even more muddled if we subdivide the data further, by geographic region. Now the formerly strong positive association between smoking and life expectancy seems very weak, if not outright nonexistent:

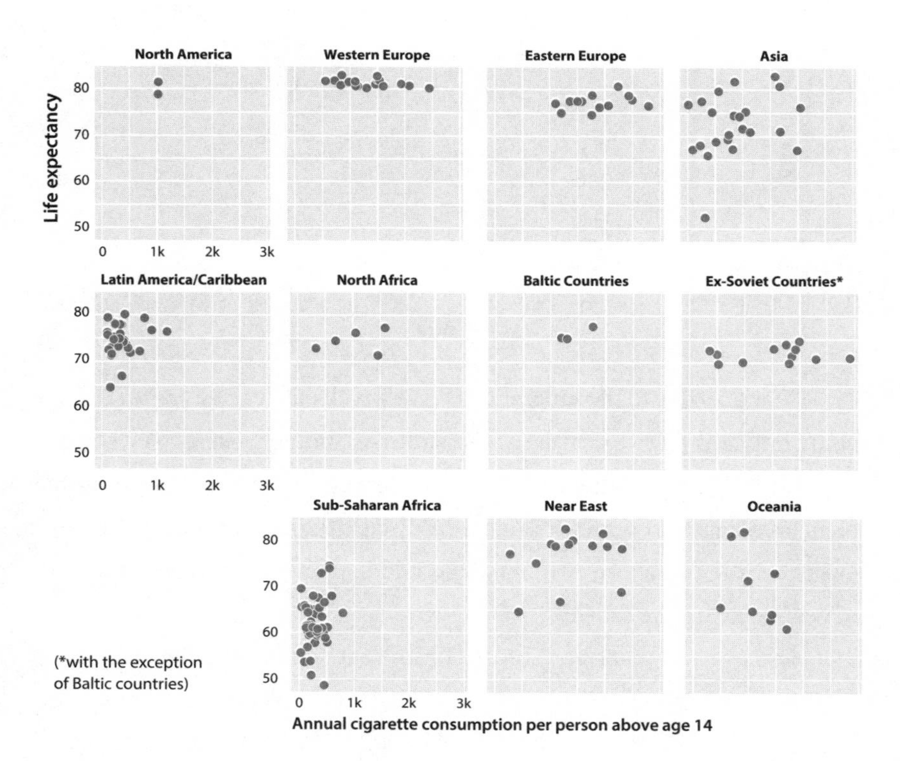

The association would vanish even further if we could split each one of those countries into its constituent parts—regions, provinces, cities, and neighborhoods—down to the individual. With each further split, the relationship between cigarette consumption and life expectancy will decrease to the point that it becomes negative: when we observe individuals, we notice that cigarette consumption has a negative effect. The following chart, based on different sources,[4] compares survival rates of people older than 40. Notice that around 50% of people who have never smoked or who stopped smoking several years ago were still alive by 80, but just a bit more than 25% of cigarette smokers survived to that point. According to several studies, cigarette smoking shortens life by around seven years (this kind of chart of survival time is called a Kaplan-Meier plot):

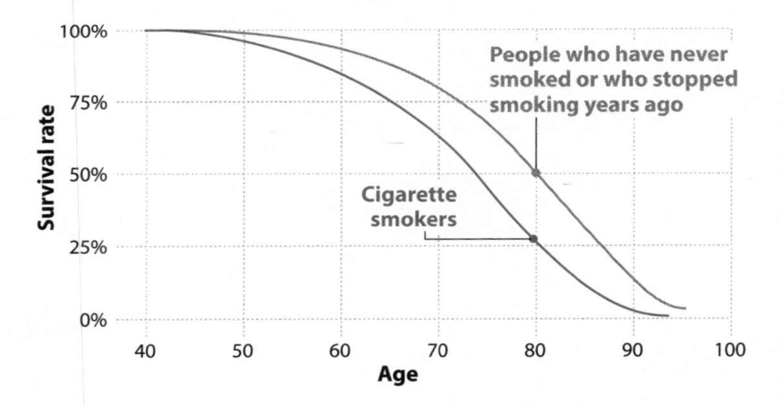

Paradoxes due to the different levels at which data can be aggregated are abundant, and they can push us into faulty reasoning. Several blog posts in the website Why Evolution Is True, by biology professor Jerry Coyne, author of an excellent book with the same title, discuss the inverse relationship that exists between religiosity on one hand and happiness and other indicators of well-being on the other.[5]

Here are two maps and a scatter plot that summarize the association between the percentage of people in each country who say that religion is important in their lives (according to a 2009 Gallup poll) and those countries' scores in the Happiness Index (a measure calculated by the United Nations for its *World Happiness Report*):

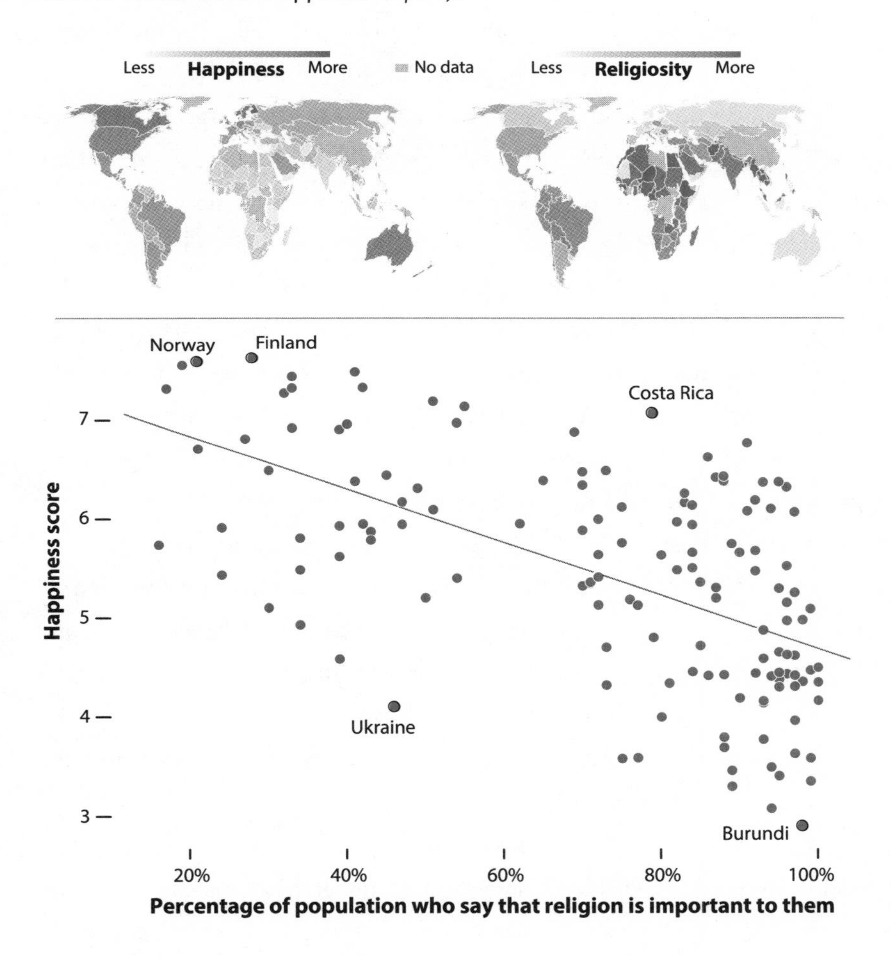

The association between the two variables is relatively weak and negative: in general, the more religious a country is, the less happy its people tend to be. The association is noticeable, even if there are plenty of exceptions. For instance, Ukraine isn't very religious, but its happiness score is low, and Costa Rica is very happy, while being very religious.

Happiness scores are positively related to equality and well-being. Countries with higher equality, as well as with well-fed and healthy inhabitants, tend to be happier. Equality and happiness are positively correlated, while equality and happiness are both inversely correlated to religiosity: the more inequality, the less happy a country tends to be and the higher the percentage of people who say that religion is important in their lives.

The inverse relationship between religiosity and indicators of happiness and well-being remains even if we disaggregate the data a bit to look at the regional level. Data from Gallup allows us to compare the percentage of people in the United States who say they are very religious with the overall well-being and life-satisfaction score of their state, a metric based on weighing factors such as access to affordable health insurance, quality of diet, amount of exercise, sense of community, and civic engagement.[6] (See the chart on the next page.) As is usual in any scatter plot, there are exceptions: West Virginia is very low in well-being and in the middle of the distribution in terms of religiosity, and Utah is high in both measures.

An enthusiastic atheist may be too quick to extract lessons from these charts. Do they mean that religiosity leads to more misery or is the opposite true? Moreover, do the charts suggest that I as an individual will be happier if I abandon my religion or even become an atheist? Of course not. Let's emphasize another rule of good chart reading:

Don't read too much into a chart—*particularly if you're reading what you'd like to read.*

First of all, these charts tell you that higher levels of religious commitment are inversely related to happiness and well-being, but they *don't* say that increasing religiosity will lead to more misery. In reality, this might be a case of reverse causality. It might be that less suffering makes countries less religious.

A study by University of Iowa professor Frederick Solt showed that year-by-year changes in inequality in different countries led to variations in religiosity, independently from how wealthy each person was in those countries. Both poor and rich people became more religious when inequality increased.[7] The rich and powerful become more religious,

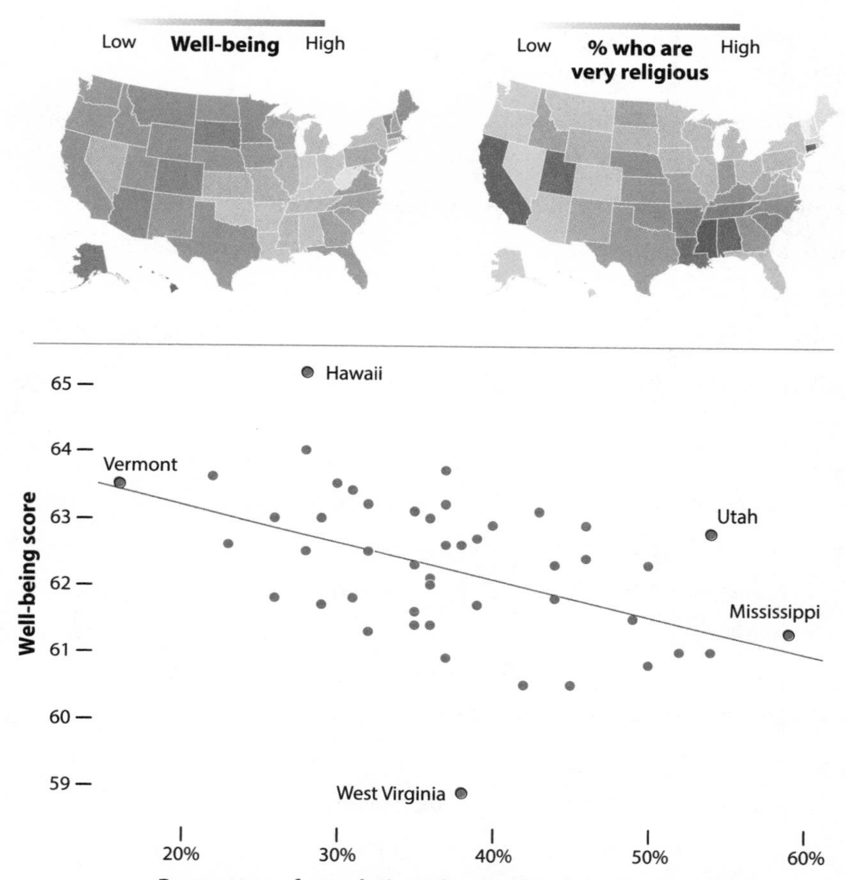

according to Solt, because religion can be used to justify societal hierarchies; to the poor, on the other hand, religion provides comfort and a sense of belonging.

This helps explain why the relationship between religiosity and happiness or perceived well-being reverses and becomes positive when we disaggregate the data even further to the individual level. This is particularly true in unstable and unequal societies, where religious people have a higher sense of well-being.[8]

Let's think of an extreme case: If you live in a poor country ravaged by war and institutional collapse, organized religion may be a strong source of meaning, consolation, community, and stability that you can latch on to. You can't compare yourself to the average Norwegian or Finnish person—who may be very happy but not very religious—and say that forfeiting religion will make you happier. Your living conditions are too different. For individuals living in rich, egalitarian, and safe places, being religious or not may not have a noticeable effect on happiness, as their societies provide health care, good education, safety, and a sense of belonging. But religion may make a huge difference for people like you. On average, being poor and religious in an unstable place may be better than being poor and nonreligious.[9]

Allow me, then, to reiterate another rule of chart reading based on the examples I've discussed so far in this chapter:

Different levels of thinking may require different levels of data aggregation.

In other words, if our goal is to learn something about the relationship between religiosity and happiness in different countries or regions, the chart should compare aggregated data of those countries or regions. If the goal is to learn about *individuals*, country-level or regional-level charts are inappropriate; instead, charts should compare people to each other.

Jumping to conclusions too quickly after we see a chart that corroborates what we already think is a universal malady that can infect anyone. After every presidential election, friends of mine who are on the left end of the political spectrum usually wonder why it often is that in poorer regions that rely more on the safety net, people tend to vote more for candidates who promise to undermine that safety net.

We could call this the "What's the Matter with Kansas?" paradox, after the title of a popular 2004 book by journalist and historian Thomas Frank. The main thesis of the book is that some voters support candidates who go against their interests, because they agree with those candidates on cultural values such as religion, abortion, gay rights, political correctness, and others. My friends are dumbfounded by charts like these:

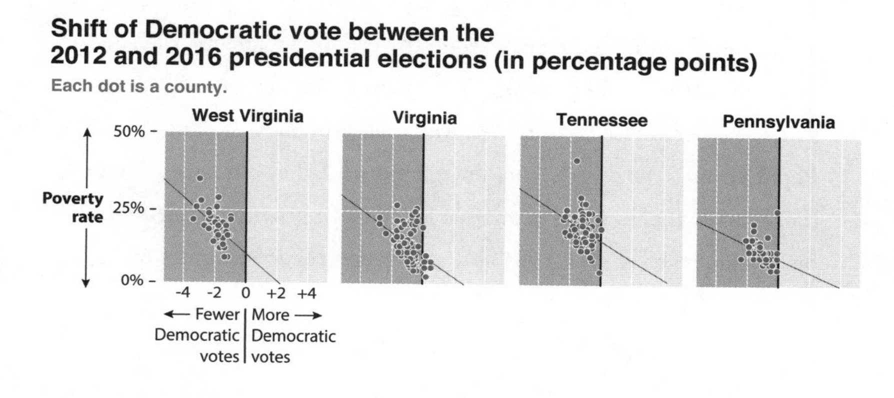

These charts seem to confirm Frank's thesis: the poorer a county is (the higher its red dot is on the chart), the more its Democratic vote decreased in 2016 in comparison to 2012 (the farther to the left its red dot is).

The pattern is real, but does it really tell us that poor people in states such as West Virginia or Tennessee are "voting against their interest"? Maybe not. To begin with, the accusation is simplistic. When we vote, we don't base our decision only on our own economic interests. I've repeatedly voted for candidates who proposed to increase taxes for families like mine. Also, voters care about candidates' values. I'd never vote for anyone who hints at anti-immigrant animus or xenophobia, no matter how much I agree with her or him on what to do with the economy.

But let's stick to the charts and imagine that economic self-interest is the only factor that voters should take into account. The charts don't become better, because what they reveal is not that poor *people* moved away from the Democratic party. They show that poorer *counties* did, which is different. Turnout in U.S. elections is usually low, and it becomes lower when you descend the economic ladder. As Alec MacGillis, a ProPublica reporter specializing in government, has written:

> The people who most rely on the safety-net programs secured by Democrats are, by and large, not voting against their own interests by electing Republicans. Rather, they are not voting, period. . . . The people in these communities who are voting Republican in larger proportions are those who are a notch or two up the economic ladder—the sheriff's deputy, the teacher, the highway worker, the motel clerk, the gas station owner and the coal miner. And their growing allegiance to the Republicans is, in part, a reaction against what they perceive, among those below them on the economic ladder, as a growing dependency on the safety net, the most visible manifestation of downward mobility in their declining towns.[10]

Thinking about differences between aggregated data and individual data is crucial to understanding how charts may bias our perceptions. Take a look at the pattern revealed by the following chart, from the website Our World in Data, which is a treasure trove if you enjoy visuals:[11]

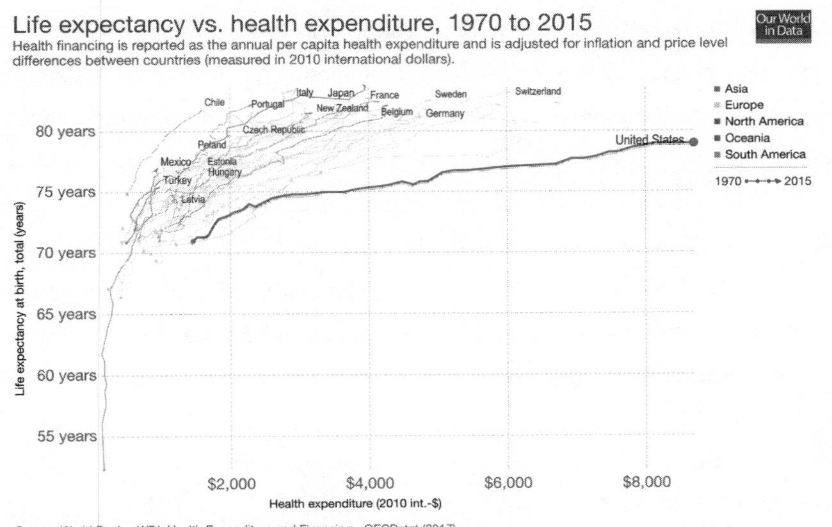

This is a connected scatter plot. We learned how to read it in chapter 2, but here's a refresher: lines correspond to countries, and you should imagine that they are like snail trails moving from left to right and from the bottom up. Focus on the United States' line. The position of the initial point, on the left, corresponds to life expectancy (vertical axis) and health expenditure per person in adjusted dollars (horizontal axis) in 1970. The end point of the U.S. line, on the right, corresponds to the same variables in 2015. This point is higher and farther to the right than the initial one, meaning that both life expectancy and health expenditures were higher in 2015 than in 1970.

What the chart reveals is that in most countries, life expectancy and health expenditures between 1970 and 2015 increased at similar rates. The exception is the United States, where life expectancy didn't increase much but average health expenditure per person soared. I can use this chart to propose another rule of good chart reading:

Any chart is a simplification of reality, and it reveals as much as it hides.

Therefore, it's always worth asking ourselves: What other patterns or trends may be hidden behind the data displayed on the chart? We could think of the variation around those national trends. Health expenditure in the United States varies enormously depending on your wealth and where you live, as does life expectancy. A 2017 study by researchers from the University of Washington found that "while residents of certain affluent counties in central Colorado had the highest life expectancy at 87 years [much more than the average Swiss or German], people in several counties of North and South Dakota, typically those with Native American reservations, could expect to die far younger, at only 66." That's a difference of more than 20 years.[12] My guess is that the variability of health expenditures and life expectancy isn't nearly as wide in wealthy nations with universal health care systems.

On March 23, 2010, President Barack Obama signed the Affordable Care Act (ACA, also known as Obamacare) into law. The ACA has been the topic of heated debates since it was proposed and up to the time of my writing in the summer of 2018. Questions include: Is it good for the economy? Is it truly affordable? Can it survive administrations that try to undermine it? Does it boost employment or does it inhibit employers from hiring?

The answers are still being debated, but some pundits have used charts like the following to argue that, contrary to what Republicans have always claimed, the ACA is really good for the job market. Notice that the number of jobs declined during the economic crisis but started recovering around 2010. Then look at what happened close to the turning point of the chart:

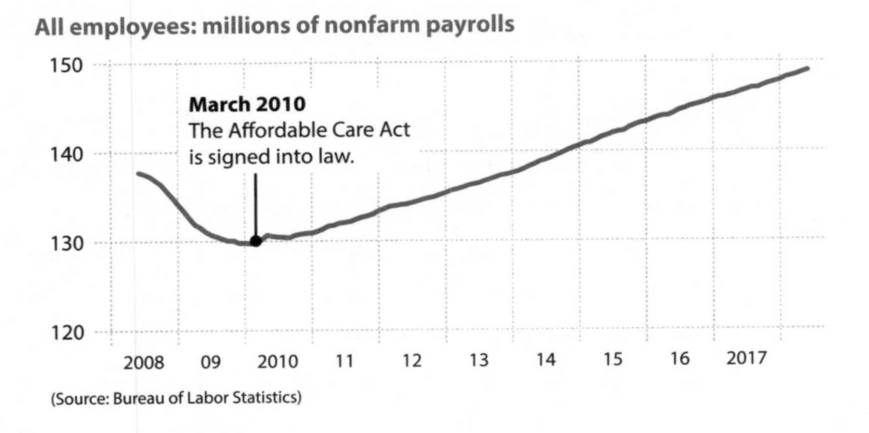

When anyone intends to persuade us with a chart, it's worth asking:

Are the patterns or trends on this chart, on their own, sufficient to support the claims the author makes?

I don't think that's the case here. The first reason is that, as we have just learned, a chart shows only what it shows, and nothing else. All this chart shows is that there are two events that happened at about the same point in time: the ACA being signed into law and the turning point of the jobs curve. But the chart doesn't say that one event caused or even influenced the other. It's your brain that makes that inference.

The second reason is that we could think of other events that also happened around those months and that could have influenced the recovery of the job market. Obama's stimulus package—the American Recovery and Reinvestment Act—was signed into law in February 2009, in response to the financial crisis of 2007–2008. It could well be that the billions of dollars injected into the economy kicked in months later, pushing companies to start hiring again.

We could also think of counterfactuals. Imagine that the Affordable Care Act had died in Congress. How would the private jobs curve change under that hypothetical? Would it be the same? Would the recovery be slower (because the ACA makes creating jobs easier) or faster (because the ACA hinders hiring, by making companies wary of health care costs)?

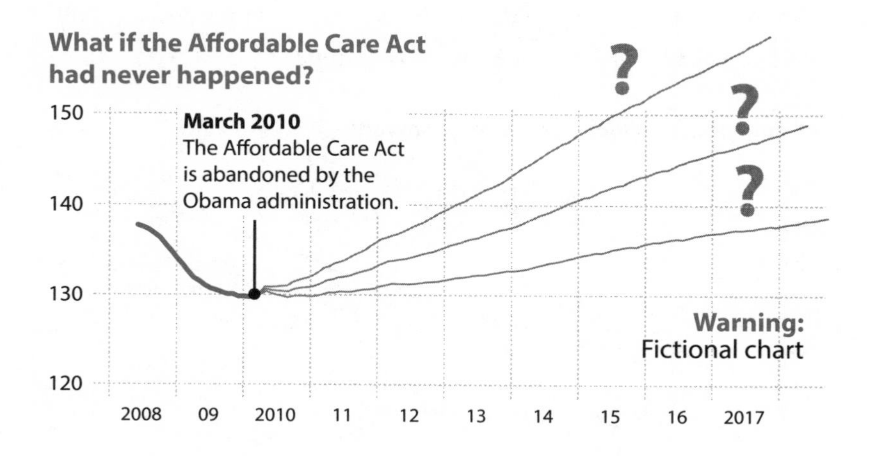

We just don't know. The original chart tells us nothing about whether the ACA had any influence on the job market. Alone, the chart is useless for either defending or attacking the program.

I've seen similar charts being mishandled by people on the right. In his first years in office, Donald Trump liked to claim that the job market was a "disaster" before he was sworn in but recovered right after, and he used charts that cropped the horizontal axis in a convenient place:

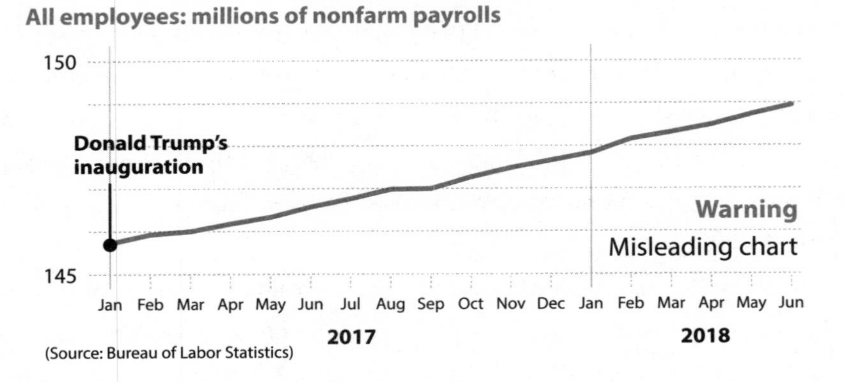

But if we go back in time and mark the point when Trump became president, we'll see that there's no remarkable change in the trajectory and slope of the line. Jobs began recovering in 2010. What Trump could have taken credit for instead was continuing an existing trend:

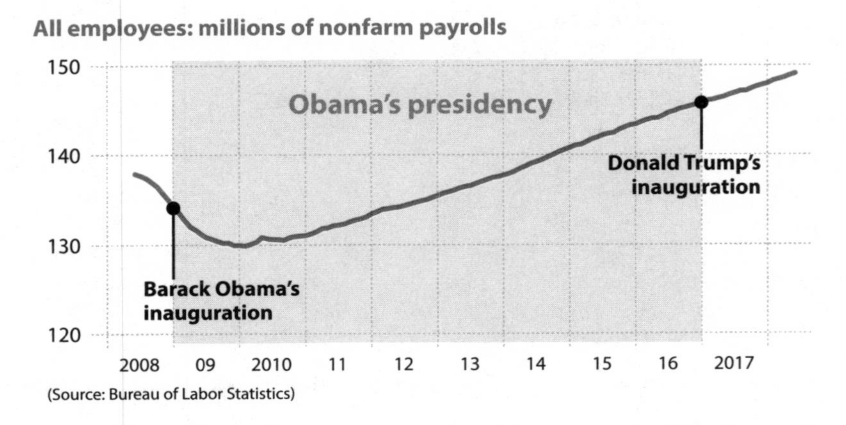

In October 2017, Trump boasted about the Dow Jones in a tweet, writing simply "WOW!" as a comment to this image revealing that the stock market was flat before Election Day in November 2016 and picked up right after:

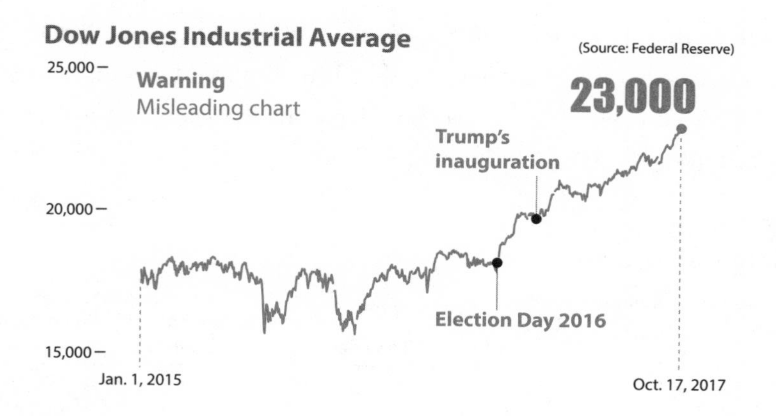

It's easy to guess what the error is: The Dow Jones follows a pattern similar to employment. It has increased quite steadily since 2009. There have been some plateaus and bumps, including the "Trump Bump" after the 2016 inauguration,[13] but the line hasn't changed direction:

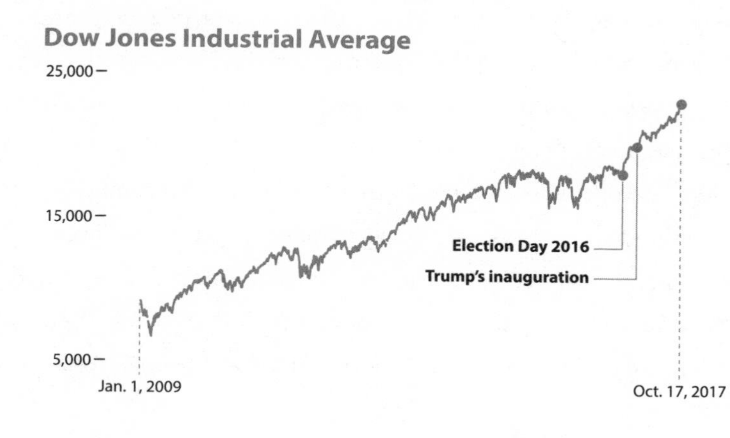

The more we cherish an idea, the more we'll love any chart that corroborates it, no matter how simplistic that chart is. The first chart below has become quite popular in creationist circles, as it reveals that there was an abrupt increase in the diversity of animal genera around a period popularly known as the Cambrian explosion. (Genera are groups of species; for instance, the genus *Canis* contains wolves, jackals, dogs, and others.) The chart is usually paired for comparison with an idealized Darwinian "tree of life," which is how evolution is supposed to work, branching out new genera steadily, little by little:

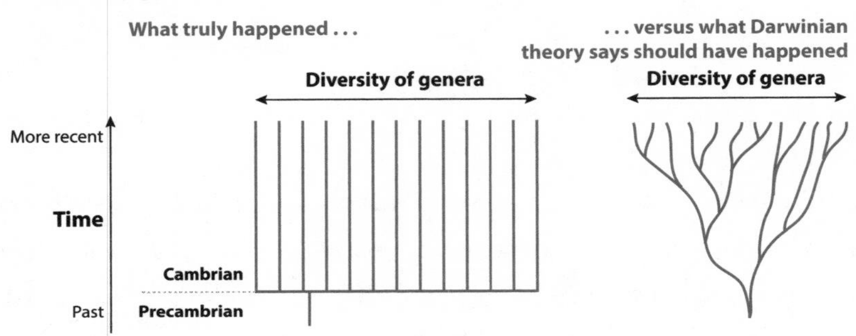

The pattern displayed on the first chart is one of sudden appearance of new kinds of animals during the Cambrian period. The Cambrian "explosion" was a mystery to biologists for more than a century—Darwin himself expressed his bafflement about it in his *On the Origin of Species*—as the incompleteness of the fossil record, particularly from Precambrian times, reinforced the idea of a rapid diversification of life. Creationists have claimed that "in a moment of geological time complex animals first appeared on earth fully formed, without evidence of any evolutionary ancestors. This remarkable explosion of life . . . is best explained by special creation by a Designer."[14]

However, the term "explosion" and the chart creationists tout are misleading. Modern scientists, benefiting from a much more complete fossil record in comparison to Darwin's time, favor the term "Cambrian diversification": Many new genera indeed appeared during the Cambrian period, but their arising was far from abrupt. The Cambrian lasted for more than 50 million years, from 545 to 490 million years ago. That's an enormous amount of time for an explosion.

Being aware of this inconvenient reality, some creationist authors such as Stephen C. Meyer stick to their chart, but they narrow the "explosion" to the third stage of the Cambrian period, the Atdabanian; occurring between 521 and 514 million years ago, this stage was when a larger diversification of genera happened. Meyer has said that "new information can only come from intelligence, and so the burst of genetic information during the Cambrian era provides compelling evidence that animal life is the product of intelligent design rather than a blind undirected process like natural selection."[15]

Seven million years is still a lot of time for a "burst"—our own species has been around for just 300,000 years—but this isn't the only problem. Donald R. Prothero, a paleontologist from Occidental College and author of the book *Evolution: What the Fossils Say and Why It Matters*, favors a much more detailed chart of the Precambrian and Cambrian periods (below) and explains that

> the entire diversification of life is now known to have gone through a number of distinct steps, from the first fossils of simple bacterial life 3.5 billion years old, to the first multicellular animals 700 m.y. ago (the Ediacara fauna), to the first evidence of skeletonized fossils (tiny fragments of small shells, nicknamed the "little shellies") at the beginning of the Cambrian, 545 m.y. ago (the Nemakit-Daldynian and Tommotian stages of the Cambrian), to the third stage of the Cambrian (Atdabanian, 520 m.y. ago), when you find the first fossils of the larger animals with hard shells, such as trilobites.[16]

Take a look at the chart below: the bars on the right represent the diversity of genera. The bars increase gradually, not suddenly. And that pattern of continuous multiplication—which ended in a mass extinction event at the end of the Botomian—began long before the Cambrian period, refuting the claim that "complex animals first appeared fully formed, without any evolutionary ancestors." You are free to believe in an "intelligent designer" if you want, but you shouldn't ignore reality.

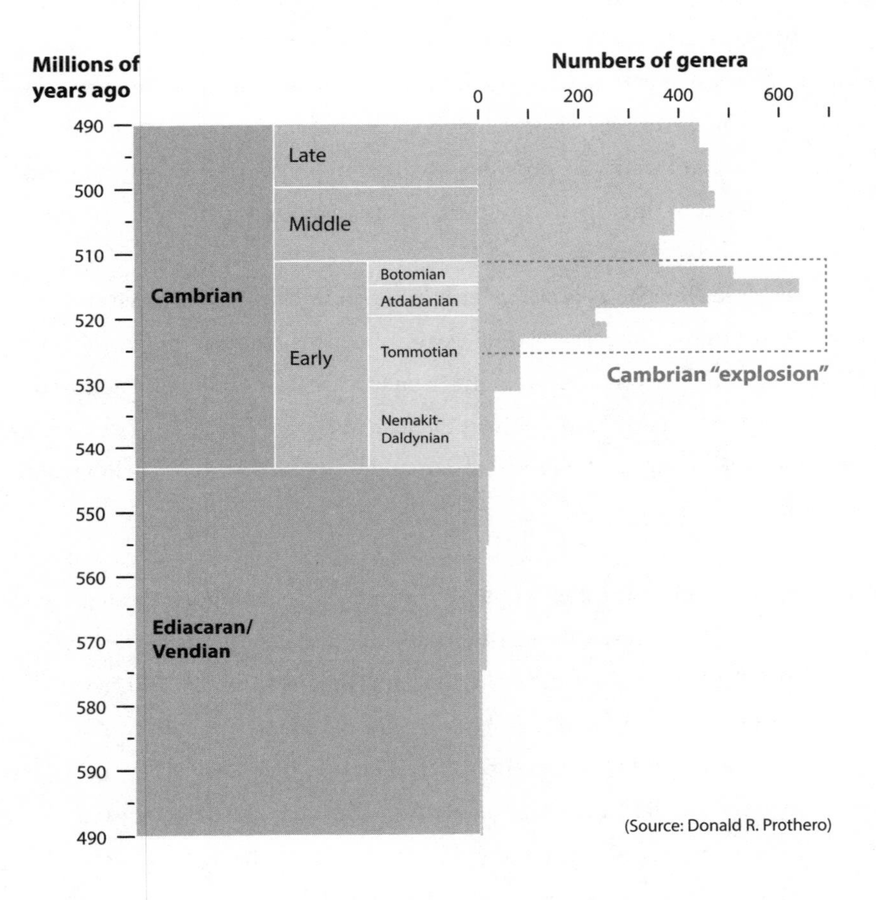

At this point, it should be clear that we can make a chart say whatever we wish—up to a point. We can do this by controlling how it is built, how much detail it contains, and, more importantly, how we interpret the patterns it displays. Read these two charts from the hilarious website Spurious Correlations, by author Tyler Vigen, who wrote a book with the same title:[17]

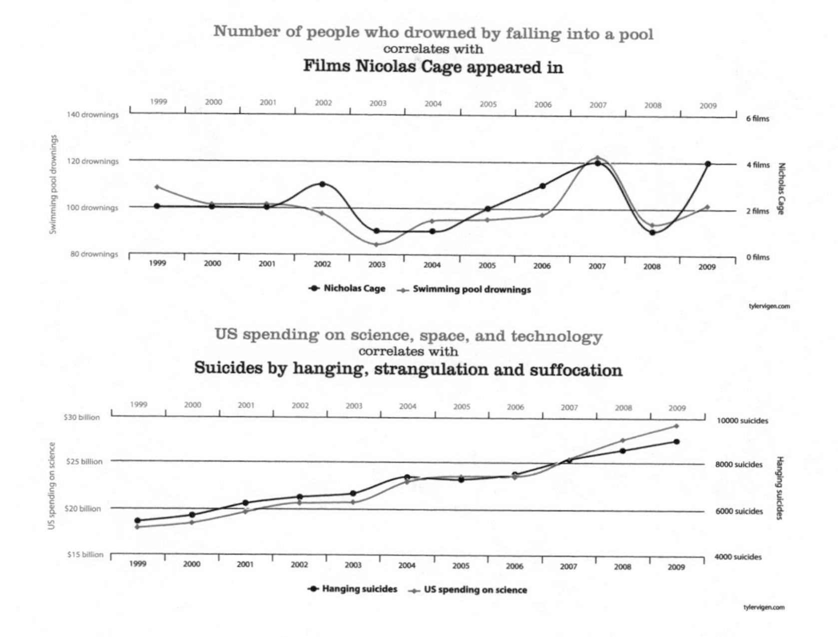

When I first visited Vigen's website, I thought that a better title—although arguably a less catchy one—would be spurious *causations*. The reason is that the number of people who drowned by falling into a pool *does* covary with the number of movies Nicolas Cage appeared in. The data is the data, and its graphical representation is fine—although the dual-axis chart can be risky sometimes; as we saw in chapter 2, we can play with the axes to make the lines adopt any slope we wish.

What is truly spurious isn't the correlation but the possible interpretation we may extract from the covariation of those variables: Does the fact that Nicolas Cage appears in more movies provoke more accidents? Or could it be that watching Nicolas Cage movies makes people more prone to swimming in their pools, exposing themselves to a higher risk of drowning? I'll let you try to come up with spurious causal links between U.S. spending in science and the number of suicides by hanging. Have fun, if you enjoy gallows humor.

Conclusion

Don't Lie to Yourself (or to Others) with Charts

If you ever visit London, after you enjoy the magnificent sights of Westminster Abbey, the Houses of Parliament, and Big Ben, cross the Westminster Bridge to the east and turn right to find St. Thomas' Hospital. There, hemmed in between large buildings, you'll find the tiny and lovely museum devoted to Florence Nightingale.

Nightingale is a beloved and debated figure in the history of public health, nursing, statistics, and chartmaking. Driven by a deep Unitarian faith, which emphasized deeds over creeds, and against the wishes of her wealthy family, Nightingale decided at an early age to devote her life to health care and to looking after the poor and needy. She also loved science. Her father had given her an excellent education in the liberal arts and mathematics. Some biographers claim that this influence helps explain why she later came to possess "one of the greatest analytical minds of her time."[1]

If you take my advice and visit the Nightingale museum in London, spend some time poring over the many documents and books showcased. One chart in particular will catch your attention:

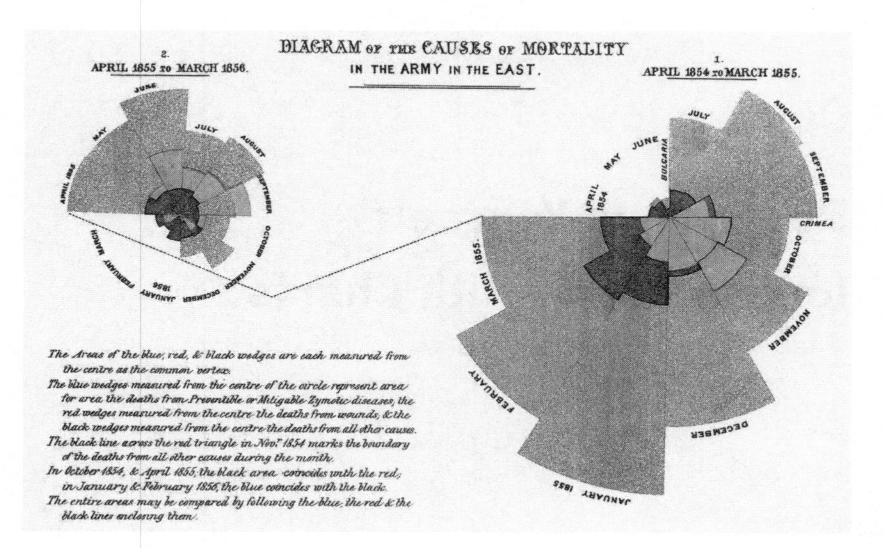

This is one of my favorite charts ever. Even if its design isn't perfect, it exemplifies many tenets of good chart reading, so allow me to give you some historical context at length.

In October of 1853, the Ottoman Empire, centered in modern Turkey, declared war on the Russian Empire. Britain and France joined what would later be called the Crimean War (1853–1856) in March 1854, on the Turkish/Ottoman side. The reasons for this war are complex and based on the expansionist desires of the Russian Empire and on disputes over the protection of Christian minorities—Russian Orthodox and Roman Catholic—who lived in Palestine, which at the time was part of the Ottoman Empire.[2]

Hundreds of thousands of soldiers perished. Mortality rates were chilling: around one out of five soldiers sent to the Crimean War died, a crushing majority of them because of diseases such as dysentery and typhoid fever, not because of wounds received during battle. At that time, there weren't effective treatments against infectious diseases—other than hydration, a good diet, and rest in a clean place—and the germ theory of disease was still two decades away.

The war was fought mostly on the Crimean Peninsula, on the northern coast of the Black Sea. British soldiers who were wounded or diagnosed with illnesses were transported to Turkey. Many died during the crossing of the Black Sea, and those who survived faced dire conditions in overcrowded, dirty, lice-ridden, and poorly supplied hospitals in Scutari (Üsküdar), part of modern Istanbul. According to researchers from Boston University, "the Scutari hospitals served more as so-called fever wards than true military hospitals and existed largely to segregate patients with fever from their healthy compatriots. Soldiers were not sent to Scutari to be healed so much as to die."[3]

Florence Nightingale, who had experience in organizing hospital supplies, volunteered to work in the Barrack Hospital of Scutari, which received that name because it was an array of repurposed military barracks. Nightingale and a team of nurses arrived in November 1854. During the almost two years she worked in Scutari, and against some resistance from the army and surgical establishments, Nightingale pushed for reforms. She kept thorough records of all patients and activities, helped improve the facilities, reduced overcrowding, pushed for more supplies, and provided psychological support to patients.

Mortality rates first increased after Nightingale's arrival, then dropped following the winter of 1854–1855, although not as fast as the legend that surrounds her often suggests. The reason, according to recent historians, is that even if Nightingale improved cleanliness, she didn't pay enough attention to ventilation or sanitation. She focused much more on the hygiene of individual patients than on the hygiene of their environment.[4]

Worried about the terrible conditions of wounded and ill soldiers, and under pressure from the opinions of a public made aware of the appalling mortality problem thanks to the press, the British government dispatched several commissions to the war zone—one to deal with supplies and another with sanitation. The Sanitary Commission began its work in March 1855. Keep this date in mind.

The Sanitary Commission, which Nightingale supported, discovered that the Barrack Hospital at Scutari sat on a cesspool, as the sewers of the building were clogged; some pipes were blocked by animal carcasses. The commissioners ordered that the sewers be cleaned, ventilation be improved, and waste be systematically disposed of. Conditions improved in all the hospitals that the commission visited, thanks to this advice.[5]

During her work at Scutari, Nightingale wasn't fully aware that the death rate at her Barrack Hospital was much higher than in other hospitals treating soldiers from the war. Some of her nurses were suspicious of the fact that amputation survival was more likely if conducted on the front than in the hospitals, but they attributed it to the fact that men on the field were "in the full vigour of life, and able to bear the pain and exhaustion, while those operated on in the hospital were worn out by suffering."[6]

To her horror, Nightingale discovered the true causes of the enormous Crimean War mortality rates only after she returned to London and analyzed the effects of the Sanitary Commission in collaboration with statisticians such as William Farr, an expert on medical hygiene. The science of hygiene was a controversial topic within the medical establishment at the time; physicians feared that their profession would be undermined if it was somehow shown that sanitation and ventilation were more important than medical care. To their dismay, this was what Nightingale's data suggested. Below you'll see a stacked bar graph—the bars add up to a total—of all deaths in the war. Notice the drop after March 1855 in overall deaths and in deaths due to diseases:

The sharp decrease in mortality can't be attributed to the improvement in hospital sanitary conditions alone, but to Nightingale it was beyond a doubt that it was a major—if not *the* major—factor.[7] Haunted by the lives that could have been saved had sanitation and ventilation been improved earlier, Nightingale devoted the remainder of her life, until she passed in 1910, to promoting the causes of nursing and public health.

This leads us back to Nightingale's chart, which she liked to call "the Wedges." When she returned from the war, Nightingale used the immense fame she had acquired to push for reform in the administration of army hospitals. She believed that the British army neglected the health and well-being of foot soldiers. The army high command disagreed, denied any responsibility, and resisted change. They had the sympathies of Queen Victoria, although she approved a royal commission to investigate the disaster in Crimea and Turkey. Nightingale contributed to this commission.

To convince the army—and, ultimately, society in general—of the wisdom of William Farr's sanitarian movement, which "urged the expenditure of public funds on flushed sewers, clean water and ventilation of dwellings,"[8] Nightingale used words, numbers, and charts that appeared not only in commission reports but also in popular books and pamphlets. The Wedges is the most famous of Nightingale's graphics. It displays the same data as the stacked bar graph, but in a much more striking and eye-catching manner.

The Wedges chart consists of two circular shapes of different sizes that are read clockwise. These shapes are made of several wedges that correspond to months. The large shape (1), to the right in the diagram below, is the period between April 1854 and March 1855, the month when the Sanitary Commission was sent to the war zone. The second, smaller shape (2), to the left, goes from April 1855 to March 1856.

There are three wedges per month, partially overlapping—but not stacking on top of—each other. The area of each wedge, measured from the center of the circle, is proportional to the deaths due to diseases, wounds, or

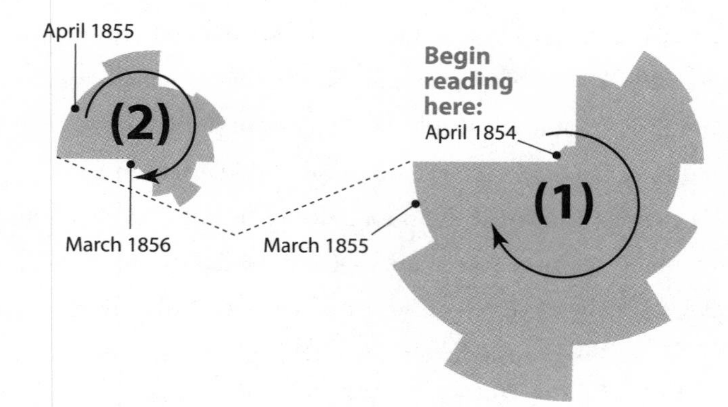

other causes, respectively. For instance, these are the segments corresponding to March 1855:

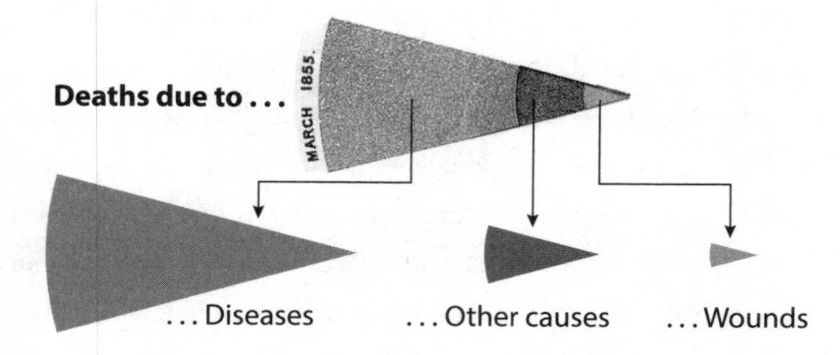

Why did Nightingale represent her data in such a fanciful way, and not as a simple stacked bar graph, or as a series of line charts, one per cause of death? Historian Hugh Small has pointed out that one of her targets was the government's chief medical officer, John Simon, who claimed that deaths due to diseases and infections were unavoidable. Nightingale wanted to prove otherwise by emphasizing the before-and-after-the-Sanitary-Commission pattern through splitting the months into two circular shapes connected with a dotted line. The first shape, showing the total deaths before the Sanitary Commission arrived, is big; the second shape is much smaller.

I'd go even further and add my own guess: I believe her goal wasn't just to inform but also to *persuade* with an intriguing, unusual, and beautiful picture. A bar graph conveys the same messages effectively, but it may not be as attractive to the eye.

The story that leads to Nightingale's famous Wedges chart embodies principles that we should all hold dear. The first and most important one, as I explained in chapter 3, is:

For a chart to be trustworthy, it must be based on reliable data.

The data Nightingale used was the best that was available at the time. It took years to collect and analyze it before it was presented to the public.

The second principle that Nightingale's Wedges exemplifies is that a chart can be a visual argument but it's rarely sufficient on its own. Nightingale's charts were always published in reports or books that explained the origins of the data and presented possible alternative interpretations before reaching a conclusion. As Karolinska Institute physician and public health statistician Hans Rosling used to say, "the world cannot be understood without numbers. And it cannot be understood with numbers alone."[9]

This thoroughness in the presentation of information is what distinguishes information from propaganda. Propaganda is information presented in a simplistic manner with the intention of shaping public opinion, highlighting what the propagandist believes strengthens his or her case and omitting what may refute it. Nightingale and her collaborators made a forceful case for public health reform, but only after painstakingly building a long and evidence-driven argument. They tried to persuade through reason.

The third principle is that data and charts can save lives and change minds. Not only other people's minds—after all, Nightingale's charts were instruments for convincing her society to change its behavior—*but your own*. This is the most compelling reason why I admire Nightingale. She was consumed by guilt after the war because the data revealed that she hadn't done enough to save the many thousands of soldiers who perished under her watch, and she acted in consequence, devoting her life to preventing future catastrophes due to the same mistakes she'd made.

The ability to change your mind in the face of evidence is, arguably, possessed by only the most honest and enlightened individuals, those who strive to use available information in the most ethical manner possible. We all ought to do our best to emulate them.

Charts can be instruments for either *reasoning* or *rationalization*. We humans are more inclined to the latter than to the former. We take the evidence that charts make visible and—particularly if we already have a belief about the topic the chart is about—we try to twist it to fit into our worldview, rather than pondering that evidence and using it to adjust our worldview accordingly.

Reasoning and rationalization rely on similar mental mechanisms. That's why they can be easily confused. They're both often based on making inferences. Inference consists of generating new information starting with available evidence or assumptions.

Inferences can be appropriate—when they correspond to reality—or not. In previous chapters, we saw a chart showing that cigarette consumption and life expectancy are positively related at a country-by-country level. We have multiple pieces of information here ("cigarette consumption, higher and lower" and "life expectancy, higher or lower"), and if we don't know better or if we have a vested interest in justifying our own cigarette consumption, we may infer that smoking cigarettes helps us live longer. Imagine that I'm a heavy smoker and am repeatedly bugged by the media, friends, and family who insist that cigarettes can kill me. If I discover a chart that suggests otherwise, I may seize on it and use it to justify my behavior. That's a rationalization.

Rationalization is the human brain's default mode. The literature about it is abundant, with dozens of popular books explaining how our mental biases lead us astray. My favorite is *Mistakes Were Made (but Not by Me)*, by Carol Tavris and Elliot Aronson. To explain how we form beliefs, justify them, and then become resistant to changing them, Tavris and Aronson talk about a metaphor they call "the pyramid of choice"; it is similar to a slippery slope.

Imagine two students with similarly mild opinions about cheating on exams. One day, during a test, both of them feel tempted to cheat. One does it, the other doesn't. Tavris and Aronson suggest that if, after this event, we could ask both students their opinions about cheating, we'll notice a significant change: the student who resisted cheating will express a much more self-righteous rejection of cheating, and the one who surrendered to the temptation will say that cheating isn't that bad a transgression or that cheating was justified in this case because a scholarship was at stake. The authors add:

> By the time the students are through with their increasingly intense levels of self-justification, two things have happened: one, they are now very far apart from one another; and two, they have internalized their beliefs and are convinced that they have always felt that way. It is as if they started off at the top of a pyramid, a millimeter apart; but by the time they have finished justifying their individual actions, they have slid to the bottom and now stand on opposite corners of its base.

Several dynamics are at work here. We humans hate dissonance. We hold ourselves in high regard and feel threatened by anything that may hurt our self-image ("I'm a good person, so it's impossible that cheating is really that bad!"). Therefore, we try to minimize threatening dissonance by rationalizing our behavior ("Everybody cheats and, besides, cheating doesn't harm anyone").

Moreover, if we later find evidence that cheating *does* indeed harm other people—if the cheater got the scholarship, then that money didn't go to a person who deserved it more—we're less likely to accept it and change our minds than we are to refuse it or twist it in a way that fits our existing belief. We behave this way because of two related human traits: the confirmation bias and motivated reasoning. Psychologist Gary Marcus wrote that "whereas confirmation bias is an automatic tendency to notice data that fit with our beliefs, motivated reasoning is the complementary tendency to scrutinize ideas more carefully if we don't like them than if we do."[10]

The relationship among cognitive dissonance, confirmation bias, and motivated reasoning is explored in books such as *The Righteous Mind* by Jonathan Haidt and *The Enigma of Reason* by Hugo Mercier and Dan Sperber. These books say that an old-fashioned—and misleading—view of human reasoning is that of a mechanism that gathers information, processes and evaluates it, and then forms beliefs based on it.

Human reason, as described by these authors, works quite differently. Reasoning, when done alone or within groups that are culturally or ideologically aligned, can decay into rationalization: we first form beliefs—because members of our groups already hold them, or because we feel emotionally good about them—and then we use our thinking skills to *justify* those beliefs, *persuade* others of their merits, and *defend* ourselves against other people's contradicting beliefs.

How do we move from rationalization to reasoning? Florence Nightingale's life leading to her charts provides useful clues. When she came back from the Crimean War, she didn't understand why so many soldiers had died under her care; she still blamed poor supplies, bureaucratic management, the poor health of the soldiers who arrived at her hospital, and so forth. She also had a reputation to maintain. Her portraits in newspapers, as a solitary nurse walking the long corridors of the Scutari hospital at night, carrying a lamp, looking after dying men, had transformed her into a popular, even mythical figure. It would have been understandable if she had succumbed to rationalizations to justify her actions during the war in Crimea.

Nightingale did the opposite: She studied her data carefully and partnered up with experts with whom she had long, heated, and honest dialogues, William Farr in particular. It was Farr who brought reams of data and evidence, and the techniques to analyze them, suggesting that improving sanitation in hospitals was the intervention that could save more lives. With Farr, Nightingale evaluated her alternative explanations for the high mortality of soldiers and weighed them against fresh numbers.

The lesson we can learn from Nightingale's experience is that, as painful as it may sound, we humans are barely able to reason on our own or when surrounded by like-minded people. When we try to reason this way, we end up rationalizing because we use arguments as self-reinforcing virtue signals. And the worst news is that the more intelligent we are and the more information we have access to, the more successful our rationalizations are. This is in part because we're more aware of what the members of the groups—political parties, churches, and others—that we belong to think, and we try to align with them. On the other hand, if you are exposed to an opinion and don't know where the opinion comes from, you're more likely to think about it on its merits.

Rationalization is a dialogue with ourselves or with like-minded brains. Reasoning, on the other hand, is an honest and open conversation in which we try to persuade interlocutors who don't necessarily agree with us beforehand with arguments that are as universally *valid*, *coherent*, and *detailed* as possible, while opening ourselves to persuasion.

This dialogue doesn't need to be face-to-face. Much dialogue in Nightingale's time happened through correspondence. When you read a paper, article, or book paying close attention, you're engaging in a dialogue with its author, the same way that if you write a book, you expect your audience to not just absorb it passively but to reflect on its contents, critique them constructively, or expand on them in the future. This is why it's so crucial to have a balanced media diet, composed of carefully curated news publications (a recommendation I made in chapter 3). Just as we're conscious of what we put inside our bodies when eating or drinking, we should exercise similar care with what we put inside our heads.

The arguments we use when we rationalize are rarely universally valid, coherent, and detailed. You can put yourself to the test. Try to explain to someone who disagrees with you on something why you believe what you believe. As much as you can, avoid arguments from authority ("This book, author, scientist, thinker—or TV anchor—said that . . .") or appeals to your values ("I'm a leftist liberal, so . . .").

Instead, lay out your case step by step, being careful to attach each link of your reasoning chain to the preceding and subsequent ones. You'll soon realize how unstable the scaffolding sustaining even our deepest and dearest beliefs is. It's a humbling experience, and it makes you realize that we should all lose our fear of admitting "I don't know." Most of the time, we really don't.

This is also one of the strategies that experts in thinking suggest to convince people who are wrong about something.[11] Don't just throw evidence at them, as this may backfire, triggering the diabolical triad of cognitive dissonance, motivated reasoning, and confirmation bias. Instead, make people deliberate. Experiments have shown that when we put people of differing opinions together in the same room and we ask them to talk to each other as equals—without identifying them as part of any kind of group, as doing so may trigger in-group defense instincts—they become more moderate. If you argue with others, show genuine interest in what they believe, empathize, and ask for a detailed explanation. Doing so may help you and them become aware of knowledge gaps. The best antidote to a misguided belief is not just truthful information. Instead, it is doubt and uncertainty, cracks in the edifice of belief through which truthful information can later leak in.

Because of their clarity and persuasiveness, charts can be key in dialogues. In a 2017 paper, political science professors Brendan Nyhan and Jason Reifler described three experiments in which charts helped dispel misperceptions.[12] The United States invaded Iraq in 2003, and in 2007 the George W. Bush administration announced a surge in the number of troops occupying the country, to deal with numerous insurgent attacks that killed both soldiers and civilians. Beginning in June that year, casualties began dropping.

Public opinion was divided about the effectiveness of the surge. According to Nyhan and Reifler, 70% of Republicans thought the surge was making the situation in Iraq better—which it was—but only 21% of Democrats agreed with them. More worrying may be the 31% of Democrats who thought that the surge was making matters even worse by helping increase violence and casualties.

Nyhan and Reifler divided the subjects in their experiments into three groups: those who wanted the U.S. to remain in Iraq, those who wanted a withdrawal, and those who didn't have a strong opinion about it. They showed them this chart:

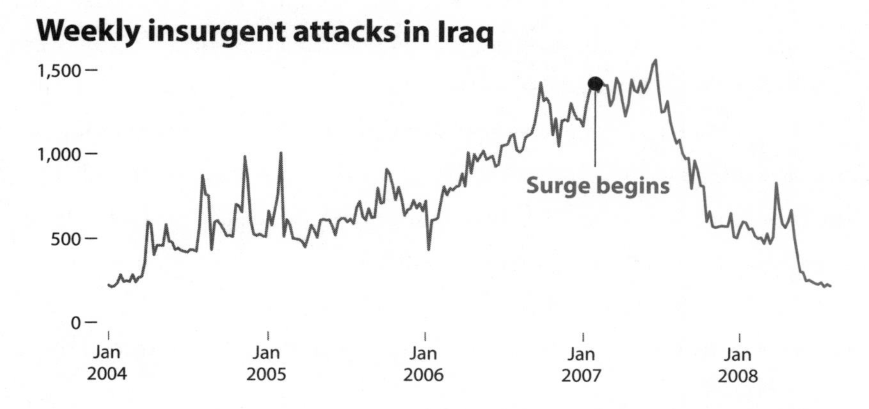

The chart reduced the proportion of people who believed that the surge had had no effect or had increased the number of attacks and victims. The effect was more visible among those who opposed the U.S. occupation of Iraq. The chart didn't change everyone's minds, but it did mitigate misperceptions in some people. Nyhan and Reifler conducted two other experiments with charts about the job market under President Obama (showing that many people, particularly Republicans, didn't believe that unemployment had dropped rapidly during his tenure) and climate change. Charts reduced—but did not eliminate—misperceptions in both cases.

Nyhan and Reifler's experiments bring me back to the main message of this book: **charts can make us smarter and can enable fruitful conversations, but only if certain conditions are met**. Some of them are related to the design of those charts; others, with the way we readers interpret them. The old saying, "There are three kinds of lies: lies, damned lies, and statistics," commonly attributed to Benjamin Disraeli and Mark Twain, is sadly popular, but statistics lie only if we want them to lie or if we lack the knowledge to make them tell the truth. Dubious charts are far more often the result of sloppiness or ignorance than of maliciousness.

Another condition is that we, as readers, should **approach charts as means to enhance conversations**. Most charts aren't conversation stoppers but conversation *enablers*. A good chart may help you answer a question ("Did the number of attacks increase or decrease after the surge?"), but charts are better at piquing our curiosity and prompting us to *ask better questions* ("But what about the number of victims?"). Think of Nightingale's case again. Her famous chart was part of a very long argument laying out the evidence that led her and her collaborators to believe that sanitation had been neglected and that something had to be done about it. But the data and the chart themselves said nothing about what could be done.

This leads us to the next condition for a chart to make us smarter: **we must stick to the principle that a chart shows only what it shows and therefore, we must strive not to read too much into it**. Nyhan and Reifler's chart shows that the number of attacks fell sharply after the surge, but it could be that each of these attacks was deadlier than each attack in the past and, as a result, that the number of victims increased. This wasn't the case, but it's a possibility, and it would be another piece of evidence we could look into to have a conversation about the consequences of the surge in Iraq.

There's another lesson I'd argue we can learn from Nightingale: **our purpose when using charts matters**.

If there's something that distinguishes the human species from other animals, it is our ability to devise technologies—both physical and conceptual—that extend our bodies and minds. We move faster thanks to wheels and wings; we see more and better thanks to glasses, telescopes, and microscopes; we enjoy deeper and more reliable memories thanks to print media and computers; we become stronger thanks to carts, cranes, and poles; and we communicate more effectively thanks to spoken and written language and the technologies intended to enable it and disseminate it. The list could go on and on, and it proves that we are a cyborg species. We can barely survive without the tools and prosthetics we imagine and realize.

Some technologies are brain prosthetics that expand our intelligence. Philosophy, logic, rhetoric, mathematics, art, and the methods of science garner our dreams, curiosity, and intuitions, channeling them in productive ways. They are conceptual tools. Charts are among them. A good chart widens our imagination and enhances our understanding by providing insights from numbers.

But tools don't just expand our bodies or perceptions. They also have an ethical dimension. Tools aren't neutral because both their design and their potential usages aren't neutral either. Tool creators have a responsibility to think about the possible consequences of the innovations they devise and to tweak their creations if those consequences turn out to be negative; on the other hand, anyone who uses a tool should try to employ it ethically. Here's a hammer:

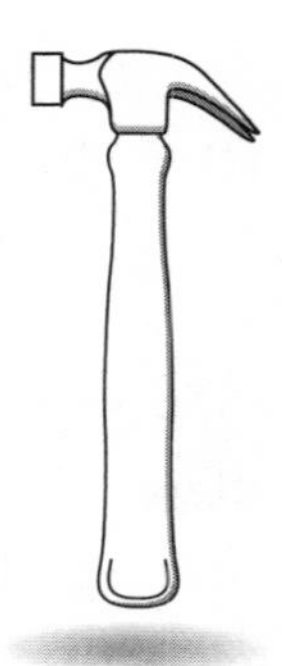

What is a hammer *for*? A hammer can be used to hit nails and build houses, shelters, barns, and walls that shield people, crops, and animals from the ravages of weather, thereby preventing misery and famine in the poorest regions of the world. In the same way, charts can be used to build understanding, communicate it to others, and inform conversations.

But the exact same hammer can be used for the opposite purpose: to destroy houses, shelters, barns, and walls, thereby condemning their owners to misery and famine. Or it can be used as a weapon of war. Similarly, charts—which are also a type of technology—can be employed to destroy understanding, mislead ourselves and others, and impede conversations.

The fight against disinformation is an endless arms race. Each generation has its new technologies and its own kind of propagandists taking advantage of them. Between the 1930s and the 1940s, the Nazis seized on technologies such as the printing press, as well as the radio and movies to promote fear, hatred, war, and genocide. If you have the opportunity, read one of the books about Nazi propaganda published by the United States Holocaust Memorial Museum,[13] or search for examples on the internet. To our modern eyes, Nazi propaganda looks strident, crude, and unpersuasive. How could people be convinced by such malarkey?

The reason is that disinformation is always as sophisticated—or as unsophisticated—as the society that engenders it. While writing these lines, I learned about a terrifying set of new artificial intelligence–driven tools that let you manipulate audio and video files.[14] You can record yourself reading a statement and then have the tool make you sound like, say, Barack Obama or Richard Nixon, given that you have recordings of their speeches that you can feed into the tool to train it. There are also technologies that can do something similar with video: record yourself making faces and then map your expressions onto another person's face.

Data and charts aren't new for scientists, mathematicians, statisticians, or engineers, but they are novel technologies in the eyes of many among the public at large, who see them as embodiments of truth. This opens the door to propagandists and liars, and the best defenses we have are education, attention, ethics, and conversations. We live in a time when data and charts are not only glorified but also ubiquitous because the means to spread them—online, particularly through social media—allow each one of us to reach dozens of people, if not hundreds, thousands, or millions.

Nearly 50,000 people follow me on Twitter. This is a sobering fact that has made me very wary of what I share on that platform. If I screw up and post something that is grossly misleading, it may rapidly be spread by many of my followers. It's happened several times, and I rushed to correct myself and contact everyone who shared what I had published.[15]

We journalists say that ours is, first, a "discipline of verification." We can always be better at adhering to that ideal, but most reporters and editors I know do take verification seriously. Perhaps it's time for this principle of verification to stop being just a journalistic ethical mandate and become instead a civic responsibility—the responsibility to assess whether what we share publicly looks and sounds right, if only to preserve the quality of our information ecosystems and public discourse. We know intuitively that we ought to use hammers responsibly—to build, not to destroy. We ought to begin thinking about other technologies such as charts and social media in the same way so instead of being part of the misinformation and disinformation malady that currently ails us, we become part of society's immune system.

In July 1982, famous evolutionary biologist and bestselling author Stephen Jay Gould was diagnosed with abdominal mesothelioma, an incurable and rare cancer caused by exposure to asbestos. His doctors informed him that the median survival time after detecting the disease was just eight months. In other words: half the patients who received diagnoses lived less than eight months, and the other half lived longer. In a marvelous essay about his experience, Gould wrote:

> Attitude clearly matters in fighting cancer. We don't know why . . . but match people with the same cancer for age, class, health, socioeconomic status, and, in general, those with positive attitudes, with a strong will and purpose for living . . . tend to live longer.[16]

But how do you develop a positive attitude when you've just discovered that people like you have, on average, just eight months to live? Perhaps you can do this by understanding that sometimes a little information is much worse than no information at all. The charts that Gould found in the medical literature probably looked similar to this fictional Kaplan-Meier chart:

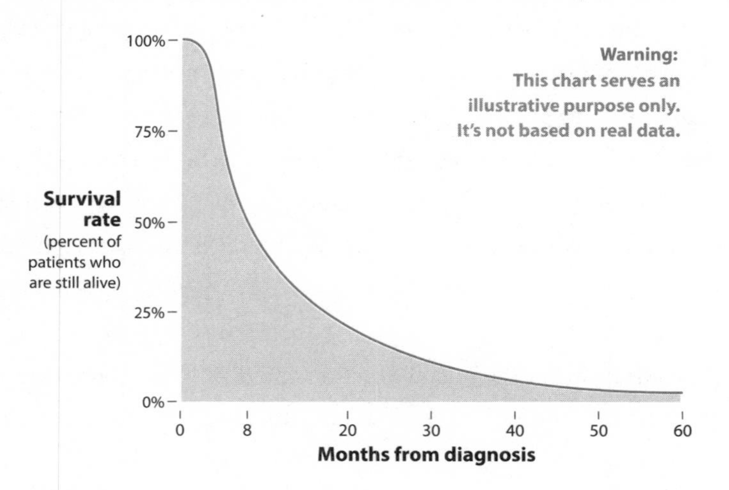

What Gould realized was that knowing that the median survival time of an abdominal mesothelioma patient was eight months didn't mean that *he* would survive that long. Charts like the one above usually display a precipitous drop in survival rates first and then a very long tail to the right.

Gould thought that he'd be on that long tail. A person's survival time after being diagnosed with any cancer depends on many factors, such as the age at which you got the bad news (Gould was relatively young), the stage of the cancer (the size of the tumor and whether it's localized or has metastasized to other parts of your body), your overall health, whether you're a smoker or not, the quality of the care and the type of treatment that you receive (Gould underwent aggressive experimental treatment), and probably your genes. Gould concluded that it was less likely that he'd be among the 50% of people who died within eight months after diagnosis than that he'd be part of the smaller percentage who went on to live for many years.

He was right. Gould was diagnosed with abdominal mesothelioma when he was 40 years old. He survived another 20 productive years that he devoted to teaching, writing dozens of popular-science articles and books, and producing a massive monograph that was published just a few months before he passed away, called *The Structure of Evolutionary Theory*.

Gould became happier, wiser, and more hopeful by meticulously assessing good numbers and charts. I dream of a future where everyone can do the same.

Acknowledgments

This book wouldn't have been possible without the support of my wife and three kids. It's been a long journey, and their presence made every day's struggles with the blank page bearable.

Many scientists and statisticians read early versions of this book and provided feedback. Nick Cox sent me a printed copy of my own first draft with thoughtful comments and corrections on every page. Diego Kuonen, Heather Krause, Frédéric Schütz, and Jon Schwabish reviewed my previous books thoroughly and did the same with this one. Other friends who made *How Charts Lie* better are John Bailer, Stephen Few, Alyssa Fowers, Kaiser Fung, Robert Grant, Ben Kirtman, Kim Kowalewski, Michael E. Mann, Alex Reinhart, Cameron Riopelle, Naomi Robbins, Walter Sosa Escudero, and Mauricio Vargas.

The School of Communication at the University of Miami, where I teach, is the best home I've had in my career. I'd like to thank its dean, Greg Shepherd, and the heads of the departments and centers I work for, Sam Terilli, Kim Grinfeder, and Nick Tsinoremas.

Besides being a professor, I have a career as a designer and consultant. I'd like to thank all my clients, particularly McMaster-Carr, Akerman, and Google News Lab's Simon Rogers and the rest of his team, for our continuous collaboration creating free chartmaking tools. Thanks also to all the educational institutions that hosted a public lecture I gave numerous times between 2017 and 2019 about the challenges I expose in *How Charts Lie*. That talk is the foundation of this book.

Some of the ideas contained in *How Charts Lie* occurred to me during conferences I've helped organize in Miami; many thanks to my partners in crime Eve Cruz, Helen Gynell, Paige Morgan, Athina Hadjixenofontos, and Greta Wells.

In chapter 5 I mentioned the cone of uncertainty. I'm currently part of a research team, led by my University of Miami colleague Barbara Millet, whose goal is to come up with better charts to inform the public about hurricane risks. Kenny Broad, Scotney Evans, and Sharan Majumdar are also part of this group. Thanks to all of them for the many fun discussions.

Finally, thanks to my agent, David Fugate, who taught me what a good book proposal looks like, and to my editor at W. W. Norton, Quynh Do, whose enthusiasm for *How Charts Lie* was a constant source of encouragement. Thank you also to W. W. Norton project editor Dassi Zeidel, copyeditor Sarah Johnson, proofreader Laura Starrett, and production manager Lauren Abbate for their excellent and careful work.

Notes

Prologue

1. I recommend David Boyle's *The Tyranny of Numbers* (London: HarperCollins, 2001).
2. I wrote about this case in one of my textbooks, *The Truthful Art: Data, Charts, and Maps for Communications* (San Francisco: New Riders, 2016).
3. Jerry Z. Muller, *The Tyranny of Metrics* (Princeton, NJ: Princeton University Press, 2018).

Introduction

1. Stephen J. Adler, Jeff Mason, and Steve Holland, "Exclusive: Trump Says He Thought Being President Would Be Easier Than His Old Life," Reuters, April 28, 2017, https://www.reuters.com/article/us-usa-trump-100days/exclusive-trump-says-he-thought-being-president-would-be-easier-than-his-old-life-idUSKBN17U0CA.
2. John Bowden, "Trump to Display Map of 2016 Election Results in the White House: Report," The Hill, November 5, 2017, http://thehill.com/blogs/blog-briefing-room/332927-trump-will-hang-map-of-2016-election-results-in-the-white-house.
3. "2016 November General Election Turnout Rates," United States Election Project, last updated September 5, 2018, http://www.electproject.org/2016g.
4. Associated Press, "Trending Story That Clinton Won Just 57 Counties Is Untrue," PBS, December 6, 2016, https://www.pbs.org/newshour/politics/trending-story-clinton-won-just-57-counties-untrue.
5. Chris Wilson, "Here's the Election Map President Trump Should Hang in the West Wing," *Time*, May 17, 2017, http://time.com/4780991/donald-trump-election-map-white-house/.
6. He announced it in a tweet: Kid Rock (@KidRock), "I have had a ton of emails and texts asking me if this website is real," Twitter, July 12, 2017, 1:51 p.m., https://

twitter.com/KidRock/status/885240249655468032. Tim Alberta and Zack Stanton, "Senator Kid Rock. Don't Laugh," Politico, July 23, 2017, https://www.politico.com/magazine/story/2017/07/23/kid-rock-run-senate-serious-michigan-analysis-215408.

7. David Weigel, "Kid Rock Says Senate 'Campaign' Was a Stunt," *Washington Post*, October 24, 2017, https://www.washingtonpost.com/news/powerpost/wp/2017/10/24/kid-rock-says-senate-campaign-was-a-stunt/?utm_term=.8d9509f4e8b4; although there's even a website: https://www.kidrockforsenate.com/.
8. Paul Krugman, "Worse Than Willie Horton," *New York Times*, January 31, 2018, https://www.nytimes.com/2018/01/31/opinion/worse-than-willie-horton.html.
9. "Uniform Crime Reporting (UCR) Program," Federal Bureau of Investigation, accessed January 27, 2019, https://ucr.fbi.gov/.
10. Richard A. Berk, professor of statistics and criminology at the University of Pennsylvania: "It isn't a national trend, it's a city trend, and it's not even a city trend, but a problem in certain neighborhoods. . . . Certainly, people around the country should not be worried. People in Chicago shouldn't be worried. But people in certain neighborhoods might be." Quoted by Timothy Williams, "Whether Crime Is Up or Down Depends on Data Being Used," *New York Times*, September 28, 2016, https://www.nytimes.com/2016/09/28/us/murder-rate-cities.html.
11. Cary Funk and Sara Kehaulani Goo, "A Look at What the Public Knows and Does Not Know about Science," Pew Research Center, September 10, 2015, http://www.pewinternet.org/2015/09/10/what-the-public-knows-and-does-not-know-about-science/.
12. Adriana Arcia et al., "Sometimes More Is More: Iterative Participatory Design of Infographics for Engagement of Community Members with Varying Levels of Health Literacy," *Journal of the American Medical Informatics Association* 23, no. 1 (January 2016): 174–83, https://doi.org/10.1093/jamia/ocv079.
13. Anshul Vikram Pandey et al., "The Persuasive Power of Data Visualization," *New York University Public Law and Legal Theory Working Papers* 474 (2014), http://lsr.nellco.org/nyu_plltwp/474.
14. The literature about cognitive biases and how they trick us is vast. I'd begin with Carol Tavris and Elliot Aronson, *Mistakes Were Made (but Not by Me): Why We Justify Foolish Beliefs, Bad Decisions, and Hurtful Acts* (New York: Mariner Books, 2007).
15. Steve King (@SteveKingIA), "Illegal immigrants are doing what Americans are reluctant to do," Twitter, February 3, 2018, 5:33 p.m., https://twitter.com/SteveKingIA/status/959963140502052867.
16. David A. Freedman, "Ecological Inference and the Ecological Fallacy," Technical Report No. 549, October 15, 1999, https://web.stanford.edu/class/ed260/freedman549.pdf.
17. W. G. V. Balchin, "Graphicacy," *Geography* 57, no. 3 (July 1972): 185–95.
18. Mark Monmonier, *Mapping It Out: Expository Cartography for the Humanities and Social Sciences* (Chicago: University of Chicago Press, 1993).
19. I give more recommendations in this book's website: http://www.howchartslie.com.

Chapter 1: How Charts Work

1. The best biography of William Playfair was written by Bruce Berkowitz, *Playfair: The True Story of the British Secret Agent Who Changed How We See the World* (Fairfax, VA: George Mason University Press, 2018).
2. Explaining how to calculate trend lines is beyond the scope of this book. For a good discussion about them and the history of the scatter plot, see Michael Friendly and Daniel Denis, "The Early Origins and Development of the Scatterplot," *Journal of the History of the Behavioral Sciences* 41, no. 2 (Spring 2005): 103–130, http://datavis.ca/papers/friendly-scat.pdf.
3. Ben Shneiderman and Catherine Plaisant, "Treemaps for Space-Constrained Visualization of Hierarchies, including the History of Treemap Research at the University of Maryland," University of Maryland, http://www.cs.umd.edu/hcil/treemap-history/.
4. Stef W. Kight, "Who Trump Attacks the Most on Twitter," *Axios*, October 14, 2017, https://www.axios.com/who-trump-attacks-the-most-on-twitter-1513305449-f084c32e-fcdf-43a3-8c55-2da84d45db34.html.
5. Stephen M. Kosslyn et al., "PowerPoint Presentation Flaws and Failures: A Psychological Analysis," *Frontiers in Psychology* 3 (2012): 230, https://www.ncbi.nlm.nih.gov/pmc/articles/PMC3398435/.
6. Matt McGrath, "China's Per Capita Carbon Emissions Overtake EU's," BBC News, September 21, 2014, http://www.bbc.com/news/science-environment-29239194.

Chapter 2: Charts That Lie by Being Poorly Designed

1. MSNBC captured the moment: TPM TV, "Planned Parenthood's Cecile Richards Shuts Down GOP Chair over Abortion Chart," YouTube, September 29, 2015, https://www.youtube.com/watch?v=iGlLLzw5_KM.
2. Linda Qiu, "Chart Shown at Planned Parenthood Hearing Is Misleading and 'Ethically Wrong,'" Politifact, October 1, 2015, http://www.politifact.com/truth-o-meter/statements/2015/oct/01/jason-chaffetz/chart-shown-planned-parenthood-hearing-misleading-/.
3. Schuch's Github repository is https://emschuch.github.io/Planned-Parenthood/ and her website is http://www.emilyschuch.com/.
4. White House Archived (@ObamaWhiteHouse), "Good news: America's high school graduation rate has increased to an all-time high," Twitter, December 16, 2015, 10:11 a.m., https://twitter.com/ObamaWhiteHouse/status/677189256834609152.
5. Keith Collins, "The Most Misleading Charts of 2015, Fixed," Quartz, December 23, 2015, https://qz.com/580859/the-most-misleading-charts-of-2015-fixed/.
6. Anshul Vikram Pandey et al., "How Deceptive Are Deceptive Visualizations? An Empirical Analysis of Common Distortion Techniques," *New York University Public Law and Legal Theory Working Papers* 504 (2015), http://lsr.nellco.org/cgi/viewcontent.cgi?article=1506&context=nyu_plltwp.
7. *National Review*'s tweet was later erased, but the *Washington Post* wrote about it:

Philip Bump, "Why this National Review global temperature graph is so misleading," December 14, 2015: https://www.washingtonpost.com/news/the-fix/wp/2015/12/14/why-the-national-reviews-global-temperature-graph-is-so-misleading/?utm_term=.dc562ee5b9f0.

8. "Federal Debt: Total Public Debt as Percent of Gross Domestic Product," FRED Economic Data, Federal Reserve Bank of St. Louis, https://fred.stlouisfed.org/series/GFDEGDQ188S.
9. Intergovernmental Panel on Climate Change, *Climate Change 2001: The Scientific Basis* (Cambridge: Cambridge University Press, 2001), https://www.ipcc.ch/ipccreports/tar/wg1/pdf/WGI_TAR_full_report.pdf.
10. Mark Monmonier himself has an entire book about this unjustly maligned projection: Mark Monmonier, *Rhumb Lines and Map Wars: A Social History of the Mercator Projection* (Chicago: University of Chicago Press, 2010).

Chapter 3: Charts That Lie by Displaying Dubious Data

1. Jakub Marian's map is here: "Number of Metal Bands Per Capita in Europe," Jakub Marian's Language Learning, Science & Art, accessed January 27, 2019, https://jakubmarian.com/number-of-metal-bands-per-capita-in-europe/. The data the map is based on can be obtained on the Encyclopaedia Metallum website, https://www.metal-archives.com/.
2. Ray Sanchez and Ed Payne, "Charleston Church Shooting: Who Is Dylann Roof?" CNN, December 16, 2016, https://www.cnn.com/2015/06/19/us/charleston-church-shooting-suspect/index.html.
3. Avalon Zoppo, "Charleston Shooter Dylann Roof Moved to Death Row in Terre Haute Federal Prison," NBC News, April 22, 2017, https://www.nbcnews.com/storyline/charleston-church-shooting/charleston-shooter-dylann-roof-moved-death-row-terre-haute-federal-n749671.
4. Rebecca Hersher, "What Happened When Dylann Roof Asked Google for Information about Race?" NPR, January 10, 2017, https://www.npr.org/sections/thetwo-way/2017/01/10/508363607/what-happened-when-dylann-roof-asked-google-for-information-about-race.
5. Jared Taylor, "DOJ: 85% of Violence Involving a Black and a White Is Black on White," Conservative Headlines, July 2015, http://conservative-headlines.com/2015/07/doj-85-of-violence-involving-a-black-and-a-white-is-black-on-white/.
6. Heather Mac Donald, "The Shameful Liberal Exploitation of the Charleston Massacre," *National Review*, July 1, 2015, https://www.nationalreview.com/2015/07/charleston-shooting-obama-race-crime/.
7. "2013 Hate Crime Statistics," Federal Bureau of Investigation, accessed January 27, 2019, https://ucr.fbi.gov/hate-crime/2013/topic-pages/incidents-and-offenses/incidentsandoffenses_final.
8. David A. Schum, *The Evidential Foundations of Probabilistic Reasoning* (Evanston, IL: Northwestern University Press, 2001).
9. The original quote was "If you torture the data enough, nature will always confess."

10. "Women Earn up to 43% Less at Barclays," BBC News, February 22, 2018, http://www.bbc.com/news/business-43156286.
11. Jeffrey A. Shaffer, "Critical Thinking in Data Analysis: The Barclays Gender Pay Gap," Data Plus Science, February 23, 2018, http://dataplusscience.com/GenderPayGap.html.
12. Sarah Cliff and Soo Oh, "America's Health Care Prices Are Out of Control. These 11 Charts Prove It," Vox, May 10, 2018, https://www.vox.com/a/health-prices.
13. You can find the reports by the International Federation of Health Plans through its website, http://www.ifhp.com. The 2015 report is available at https://fortunedotcom.files.wordpress.com/2018/04/66c7d-2015comparativepricereport09-09-16.pdf.
14. To learn more about the different kinds of random sampling, see this short online introduction: "Sampling," Yale University, accessed January 27, 2019, http://www.stat.yale.edu/Courses/1997-98/101/sample.htm.
15. The source for the graphic is Christopher Ingraham, "Kansas Is the Nation's Porn Capital, according to Pornhub," *WonkViz* (blog), accessed January 27, 2019, http://wonkviz.tumblr.com/post/82488570278/kansas-is-the-nations-porn-capital-according-to. He used data from Pornhub, which teamed up with BuzzFeed: Ryan Broderick, "Who Watches More Porn: Republicans or Democrats?" BuzzFeed News, April 11, 2014, https://www.buzzfeednews.com/article/ryanhatesthis/who-watches-more-porn-republicans-or-democrats.
16. Benjamin Edelman, "Red Light States: Who Buys Online Adult Entertainment?" *Journal of Economic Perspectives* 23, no. 1 (2009): 209–220, http://people.hbs.edu/bedelman/papers/redlightstates.pdf.
17. Eric Black, "Carl Bernstein Makes the Case for 'the Best Obtainable Version of the Truth,'" *Minneapolis Post*, April 17, 2015, https://www.minnpost.com/eric-black-ink/2015/04/carl-bernstein-makes-case-best-obtainable-version-truth.
18. See Tom Nichols, *The Death of Expertise: The Campaign against Established Knowledge and Why It Matters* (New York: Oxford University Press, 2017).

Chapter 4: Charts That Lie by Displaying Insufficient Data

1. "It's Time to End Chain Migration," The White House, December 15, 2017, https://www.whitehouse.gov/articles/time-end-chain-migration/.
2. Michael Shermer, *The Believing Brain: From Ghosts and Gods to Politics and Conspiracies—How We Construct Beliefs and Reinforce Them as Truths* (New York: Times Books, Henry Holt, 2011).
3. John Binder, "2,139 DACA Recipients Convicted or Accused of Crimes against Americans," Breitbart, September 5, 2017, http://www.breitbart.com/big-government/2017/09/05/2139-daca-recipients-convicted-or-accused-of-crimes-against-americans/.
4. Miriam Valverde, "What Have Courts Said about the Constitutionality of DACA?" PolitiFact, September 11, 2017, http://www.politifact.com/truth-o-meter/statements/2017/sep/11/eric-schneiderman/has-daca-been-ruled-unconstitutional/.
5. Sarah K. S. Shannon et al., "The Growth, Scope, and Spatial Distribution of People with Felony Records in the United States, 1948 to 2010," *Demography* 54, no. 5 (2017): 1795–1818, http://users.soc.umn.edu/~uggen/former_felons_2016.pdf.

6. "Family Income in 2017," FINC-01. Selected Characteristics of Families by Total Money Income, United States Census Bureau, accessed January 27, 2019, https://www.census.gov/data/tables/time-series/demo/income-poverty/cps-finc/finc-01.html.
7. TPC Staff, "Distributional Analysis of the Conference Agreement for the Tax Cuts and Jobs Act," Tax Policy Center, December 18, 2017, https://www.taxpolicycenter.org/publications/distributional-analysis-conference-agreement-tax-cuts-and-jobs-act.
8. The story is even more complicated. According to many projections, plenty of families will end up paying more taxes, not less: Danielle Kurtzleben, "Here's How GOP's Tax Breaks Would Shift Money to Rich, Poor Americans," NPR, November 14, 2017, https://www.npr.org/2017/11/14/562884070/charts-heres-how-gop-s-tax-breaks-would-shift-money-to-rich-poor-americans. Also, PolitiFact has criticized Ryan's numbers: Louis Jacobson, "Would the House GOP Tax Plan Save a Typical Family $1,182?" PolitiFact, November 3, 2017, http://www.politifact.com/truth-o-meter/statements/2017/nov/03/paul-ryan/would-house-gop-tax-plan-save-typical-family-1182/.
9. Alissa Wilkinson, "Black Panther Just Keeps Smashing Box Office Records," Vox, April 20, 2018, https://www.vox.com/culture/2018/4/20/17261614/black-panther-box-office-records-gross-iron-man-thor-captain-america-avengers.
10. Box Office Mojo has rankings of the highest-grossing movies worldwide after adjusting for inflation. *Black Panther* is the 30th: "All Time Box Office," Box Office Mojo, accessed January 27, 2019, https://www.boxofficemojo.com/alltime/adjusted.htm.
11. Rody's website is Data + Tableau + Me, http://www.datatableauandme.com.
12. "CPI Inflation Calculator," Bureau of Labor Statistics, accessed January 27, 2019, https://www.bls.gov/data/inflation_calculator.htm.
13. Dawn C. Chmielewski, "Disney Expects $200-Million Loss on 'John Carter,'" *Los Angeles Times*, March 20, 2012, http://articles.latimes.com/2012/mar/20/business/la-fi-ct-disney-write-down-20120320.
14. "Movie Budget and Financial Performance Records," The Numbers, accessed January 27, 2019, https://www.the-numbers.com/movie/budgets/.
15. "The 17 Goals," The Global Goals for Sustainable Development, accessed January 27, 2019, https://www.globalgoals.org/.
16. Defend Assange Campaign (@DefendAssange), Twitter, September 2, 2017, 8:41 a.m., https://twitter.com/julianassange/status/904006478616551425?lang=en.

Chapter 5: Charts That Lie by Concealing or Confusing Uncertainty

1. Bret Stephens, "Climate of Complete Certainty," *New York Times*, April 28, 2017, https://www.nytimes.com/2017/04/28/opinion/climate-of-complete-certainty.html.
2. Shaun A. Marcott et al., "A Reconstruction of Regional and Global Temperature for the Past 11,300 Years," *Science* 339 (2013): 1198, http://content.csbs.utah.edu/~mli/Economics%207004/Marcott_Global%20Temperature%20Reconstructed.pdf.
3. There's a book that describes how the "hockey stick" was designed: Michael E.

Mann, *The Hockey Stick and the Climate Wars: Dispatches from the Front Lines* (New York: Columbia University Press, 2012).

4. I. Allison et al., *The Copenhagen Diagnosis, 2009: Updating the World on the Latest Climate Science* (Sydney, Australia: University of New South Wales Climate Change Research Centre, 2009).
5. Heather's blog is a marvelous read if you want to learn to reason about numbers: Heather Krause, *Datablog*, https://idatassist.com/datablog/.
6. Kenny has written extensively about how the public misconstrues storm maps and graphs. For instance: Kenneth Broad et al., "Misinterpretations of the 'Cone of Uncertainty' in Florida during the 2004 Hurricane Season," *Bulletin of the American Meteorological Society* (May 2007): 651–68, https://journals.ametsoc.org/doi/pdf/10.1175/BAMS-88-5-651.
7. National Hurricane Center, "Potential Storm Surge Flooding Map," https://www.nhc.noaa.gov/surge/inundation/.

Chapter 6: Charts That Lie by Suggesting Misleading Patterns

1. From John W. Tukey, *Exploratory Data Analysis* (Reading, MA: Addison-Wesley, 1977).
2. For more details about this case, read Heather Krause, "Do You Really Know How to Use Data Correctly?" DataAssist, May 16, 2018, https://idatassist.com/do-you-really-know-how-to-use-data-correctly/.
3. The most famous amalgamation paradox is Simpson's paradox: Wikipedia, s.v. "Simpson's Paradox," last edited January 23, 2019, https://en.wikipedia.org/wiki/Simpson%27s_paradox.
4. Numerous studies display similar survival curves. See, for instance, Richard Doll et al., "Mortality in Relation to Smoking: 50 Years' Observations on Male British Doctors," *BMJ* 328 (2004): 1519, https://www.bmj.com/content/328/7455/1519.
5. Jerry Coyne, "The 2018 UN World Happiness Report: Most Atheistic (and Socially Well Off) Countries Are the Happiest, While Religious Countries Are Poor and Unhappy," Why Evolution Is True (March 20, 2018), https://whyevolutionistrue.wordpress.com/2018/03/20/the-2018-un-world-happiness-report-most-atheistic-and-socially-well-off-countries-are-the-happiest-while-religious-countries-are-poor-and-unhappy/.
6. "State of the States," Gallup, accessed January 27, 2019, https://news.gallup.com/poll/125066/State-States.aspx.
7. Frederick Solt, Philip Habel, and J. Tobin Grant, "Economic Inequality, Relative Power, and Religiosity," *Social Science Quarterly* 92, no. 2: 447–65, https://onlinelibrary.wiley.com/doi/pdf/10.1111/j.1540-6237.2011.00777.x.
8. Nigel Barber, "Are Religious People Happier?" *Psychology Today*, November 20, 2012, https://www.psychologytoday.com/us/blog/the-human-beast/201211/are-religious-people-happier.
9. Sally Quinn, "Religion Is a Sure Route to True Happiness," *Washington Post*, January 24, 2014, https://www.washingtonpost.com/national/religion/religion-is-a-sure

-route-to-true-happiness/2014/01/23/f6522120-8452-11e3-bbe5-6a2a3141e3a9_story .html?utm_term=.af77dde8deac.

10. Alec MacGillis, "Who Turned My Blue State Red?" *New York Times*, November 22, 2015, https://www.nytimes.com/2015/11/22/opinion/sunday/who-turned-my-blue-state-red .html.
11. Our World in Data (website), Max Roser, accessed January 27, 2019, https://ourworldindata.org/.
12. Richard Luscombe, "Life Expectancy Gap between Rich and Poor US Regions Is 'More Than 20 Years,'" May 8, 2017, *Guardian*, https://www.theguardian.com/inequality/2017/may/08/life-expectancy-gap-rich-poor-us-regions-more-than-20-years.
13. Harold Clarke, Marianne Stewart, and Paul Whiteley, "The 'Trump Bump' in the Stock Market Is Real. But It's Not Helping Trump," *Washington Post*, January 9, 2018, https://www.washingtonpost.com/news/monkey-cage/wp/2018/01/09/the-trump-bump-in -the-stock-market-is-real-but-its-not-helping-trump/?utm_term=.109918a60cba.
14. Description of the documentary *Darwin's Dilemma: The Mystery of the Cambrian Fossil Record*, by Stand to Reason: https://store.str.org/ProductDetails.asp?ProductCode=DVD018
15. Stephen C. Meyer, *Darwin's Doubt: The Explosive Origin of Animal Life and the Case for Intelligent Design* (New York: HarperOne, 2013).
16. Daniel R. Prothero, *Evolution: What the Fossils Say and Why It Matters* (New York: Columbia University Press, 2007).
17. http://www.tylervigen.com/spurious-correlations.

Conclusion: Don't Lie to Yourself (or to Others) with Charts

1. Mark Bostridge, *Florence Nightingale: The Woman and Her Legend* (London: Penguin Books, 2008).
2. *Encyclopaedia Britannica Online*, s.v. "Crimean War," November 27, 2018, https://www.britannica.com/event/Crimean-War.
3. Christopher J. Gill and Gillian C. Gill, "Nightingale in Scutari: Her Legacy Reexamined," *Clinical Infectious Diseases* 40, no. 12 (June 15, 2005): 1799–1805, https://doi .org/10.1086/430380.
4. Hugh Small, *Florence Nightingale: Avenging Angel* (London: Constable, 1998).
5. Bostridge, Florence Nightingale.
6. Hugh Small, *A Brief History of Florence Nightingale: And Her Real Legacy, a Revolution in Public Health* (London: Constable, 2017).
7. The data can be found here: "Mathematics of the Coxcombs," Understanding Uncertainty, May 11, 2008, https://understandinguncertainty.org/node/214.
8. Small, *Florence Nightingale.*
9. Hans Rosling, Anna Rosling Rönnlund, and Ola Rosling, *Factfulness: Ten Reasons We're Wrong about the World—And Why Things Are Better Than You Think* (New York: Flatiron Books, 2018).
10. Gary Marcus, Kluge: *The Haphazard Evolution of the Human Mind* (Boston: Mariner Books, 2008).

11. Steven Sloman and Philip Fernbach, *The Knowledge Illusion* (New York: Riverhead Books, 2017). This is the best book I've read about these matters.
12. Brendan Nyhan and Jason Reifler, "The Role of Information Deficits and Identity Threat in the Prevalence of Misperceptions," (forthcoming, *Journal of Elections, Public Opinion and Parties*, published ahead of print May 6, 2018, https://www.tandfonline.com/eprint/PCDgEX8KnPVYyytUyzvy/full).
13. For instance, Susan Bachrach and Steven Luckert, *State of Deception: The Power of Nazi Propaganda* (New York: W. W. Norton, 2009).
14. Heather Bryant, "The Universe of People Trying to Deceive Journalists Keeps Expanding, and Newsrooms Aren't Ready," http://www.niemanlab.org/2018/07/the-universe-of-people-trying-to-deceive-journalists-keeps-expanding-and-newsrooms-arent-ready/.
15. I explained the screwup in my personal blog, *The Functional Art*: http://www.thefunctionalart.com/2014/05/i-should-know-better-journalism-is.html.
16. Stephen Jay Gould, "The Median Isn't the Message," CancerGuide, last updated May 31, 2002, https://www.cancerguide.org/median_not_msg.html.

Bibliography

Bachrach, Susan, and Steven Luckert. *State of Deception: The Power of Nazi Propaganda.* New York: W. W. Norton, 2009.

Berkowitz, Bruce. *Playfair: The True Story of the British Secret Agent Who Changed How We See the World.* Fairfax, VA: George Mason University Press, 2018.

Bertin, Jacques. *Semiology of Graphics: Diagrams, Networks, Maps.* Redlands, CA: ESRI Press, 2011.

Börner, Katy. *Atlas of Knowledge: Anyone Can Map.* Cambridge, MA: MIT Press, 2015.

Bostridge, Mark. *Florence Nightingale: The Woman and Her Legend.* London: Penguin Books, 2008.

Boyle, David. *The Tyranny of Numbers.* London: HarperCollins, 2001.

Cairo, Alberto. *The Truthful Art: Data, Charts, and Maps for Communication.* San Francisco: New Riders, 2016.

Caldwell, Sally. *Statistics Unplugged.* 4th ed. Belmont, CA: Wadsworth Cengage Learning, 2013.

Card, Stuart K., Jock Mackinlay, and Ben Shneiderman. *Readings in Information Visualization: Using Vision to Think.* San Francisco: Morgan Kaufmann, 1999.

Cleveland, William. *The Elements of Graphing Data.* 2nd ed. Summit, NJ: Hobart Press, 1994.

Coyne, Jerry. *Why Evolution Is True.* New York: Oxford University Press, 2009.

Deutsch, David. *The Beginning of Infinity: Explanations That Transform the World.* New York: Viking, 2011.

Ellenberg, Jordan. *How Not to Be Wrong: The Power of Mathematical Thinking.* New York: Penguin Books, 2014.

Few, Stephen. *Show Me the Numbers: Designing Tables and Graphs to Enlighten.* 2nd ed. El Dorado Hills, CA: Analytics Press, 2012.

Fung, Kaiser. *Numbersense: How to Use Big Data to Your Advantage.* New York: McGraw Hill, 2013.

Gigerenzer, Gerd. *Calculated Risks: How to Know When Numbers Deceive You*. New York: Simon and Schuster, 2002.

Goldacre, Ben. *Bad Science: Quacks, Hacks, and Big Pharma Flacks*. New York: Farrar, Straus and Giroux, 2010.

Haidt, Jonathan. *The Righteous Mind: Why Good People Are Divided by Politics and Religion*. New York: Vintage Books, 2012.

Huff, Darrell. *How to Lie with Statistics*. New York: W. W. Norton, 1993.

Kirk, Andy. *Data Visualisation: A Handbook for Data Driven Design*. Los Angeles: Sage, 2016.

MacEachren, Alan M. *How Maps Work: Representation, Visualization, and Design*. New York: Guilford Press, 2004.

Malamed, Connie. *Visual Language for Designers: Principles for Creating Graphics That People Understand*. Beverly, MA: Rockport Publishers, 2011.

Mann, Michael E. *The Hockey Stick and the Climate Wars: Dispatches from the Front Lines*. New York: Columbia University Press, 2012.

Marcus, Gary. *Kluge: The Haphazard Evolution of the Human Mind*. Boston: Mariner Books, 2008.

Meirelles, Isabel. *Design for Information: An Introduction to the Histories, Theories, and Best Practices behind Effective Information Visualizations*. Beverly, MA: Rockport Publishers, 2013.

Mercier, Hugo, and Dan Sperber. *The Enigma of Reason*. Cambridge, MA: Harvard University Press, 2017.

Monmonier, Mark. *How to Lie with Maps*. 2nd ed. Chicago: University of Chicago Press, 2014.

———. *Mapping It Out: Expository Cartography for the Humanities and Social Sciences*. Chicago: University of Chicago Press, 1993.

Muller, Jerry Z. *The Tyranny of Metrics*. Princeton, NJ: Princeton University Press, 2018.

Munzner, Tamara. *Visualization Analysis and Design*. Boca Raton, FL: CRC Press, 2015.

Nichols, Tom. *The Death of Expertise: The Campaign against Established Knowledge and Why It Matters*. New York: Oxford University Press, 2017.

Nussbaumer Knaflic, Cole. *Storytelling with Data: A Data Visualization Guide for Business Professionals*. Hoboken, NJ: John Wiley and Sons, 2015.

Pearl, Judea, and Dana Mackenzie. *The Book of Why: The New Science of Cause and Effect*. New York: Basic Books, 2018.

Pinker, Steven. *Enlightenment Now: The Case for Reason, Science, Humanism, and Progress*. New York: Viking, 2018.

Prothero, Donald R. *Evolution: What the Fossils Say and Why It Matters*. New York: Columbia University Press, 2007.

Rosling, Hans, Anna Rosling Rönnlund, and Ola Rosling. *Factfulness: Ten Reasons We're Wrong About the World: And Why Things Are Better Than You Think*. New York: Flatiron Books, 2018.

Silver, Nate. *The Signal and the Noise: Why So Many Predictions Fail—but Some Don't*. New York: Penguin Books, 2012.

Schum, David A. *The Evidential Foundations of Probabilistic Reasoning*. Evanston, IL: Northwestern University Press, 2001.

Shermer, Michael. *The Believing Brain: From Ghosts and Gods to Politics and Conspiracies:*

How We Construct Beliefs and Reinforce Them as Truths. New York: Times Books / Henry Holt, 2011.

Sloman, Steven, and Philip Fernbach. *The Knowledge Illusion: Why We Never Think Alone*. New York: Riverhead Books, 2017.

Small, Hugh. *A Brief History of Florence Nightingale: And Her Real Legacy, a Revolution in Public Health*. London: Constable, 2017.

———. *Florence Nightingale: Avenging Angel*. London: Constable, 1998.

Tavris, Carol, and Elliot Aronson. *Mistakes Were Made (but Not by Me): Why We Justify Foolish Beliefs, Bad Decisions, and Hurtful Acts*. Boston: Houghton Mifflin Harcourt, 2007.

Tukey, John W. *Exploratory Data Analysis*. Reading, MA: Addison-Wesley, 1977.

Wainer, Howard. *Visual Revelations: Graphical Tales of Fate and Deception From Napoleon Bonaparte to Ross Perot*. London, UK: Psychology Press, 2000.

Ware, Colin. *Information Visualization: Perception for Design*. 3rd ed. Waltham, MA: Morgan Kaufmann, 2013.

Wheelan, Charles. *Naked Statistics: Stripping the Dread from the Data*. New York: W. W. Norton, 2013.

Wilkinson, Leland. *The Grammar of Graphics*. 2nd ed. New York: Springer, 2005.

Wong, Dona M. *The Wall Street Journal Guide to Information Graphics: The Dos and Don'ts of Presenting Data, Facts, and Figures*. New York: W. W. Norton, 2013.

Further Reading

In more than two decades designing charts and teaching how to make them I've realized that becoming a good graphics reader doesn't just depend on understanding their symbols and grammar. It also requires grasping the power and limitations of the numbers they depict and being mindful of how our brains deceive themselves below our awareness. Numerical literacy (numeracy) and graphical literacy (graphicacy) are connected, and they are inseparable from psychological literacy, for which we lack a good name.

If *How Charts Lie* has gotten you interested in numeracy, graphicacy, and the limitations of human reasoning, here are some readings that are natural follow-ups.

Books about reasoning:

- Tavris, Carol, and Elliot Aronson. *Mistakes Were Made (but Not by Me): Why We Justify Foolish Beliefs, Bad Decisions, and Hurtful Acts.* Boston: Houghton Mifflin Harcourt, 2007.
- Haidt, Jonathan. *The Righteous Mind: Why Good People Are Divided by Politics and Religion.* New York: Vintage Books, 2012.

- Mercier, Hugo, and Dan Sperber. *The Enigma of Reason*. Cambridge, MA: Harvard University Press, 2017.

Books about numeracy:

- Goldacre, Ben. *Bad Science: Quacks, Hacks, and Big Pharma Flacks*. New York: Farrar, Straus and Giroux, 2010.
- Wheelan, Charles. *Naked Statistics: Stripping the Dread from the Data*. New York: W. W. Norton, 2013.
- Ellenberg, Jordan. *How Not to Be Wrong: The Power of Mathematical Thinking*. New York: Penguin Books, 2014.
- Silver, Nate. *The Signal and the Noise: Why So Many Predictions Fail—but Some Don't*. New York: Penguin Books, 2012.

Books about charts:

- Wainer, Howard. *Visual Revelations: Graphical Tales of Fate and Deception From Napoleon Bonaparte To Ross Perot*. London, UK: Psychology Press, 2000.

  Wainer has many other relevant books, and has written extensively about how charts mislead us.
- Meirelles, Isabel. *Design for Information: An Introduction to the Histories, Theories, and Best Practices behind Effective Information Visualizations*. Beverly, MA: Rockport Publishers, 2013.
- Nussbaumer Knaflic, Cole. *Storytelling with Data: A Data Visualization Guide for Business Professionals*. Hoboken, NJ: John Wiley and Sons, 2015.
- Monmonier, Mark. *How to Lie with Maps*. 2nd ed. Chicago: University of Chicago Press, 2014.
- Few, Stephen. *Show Me the Numbers: Designing Tables and Graphs to Enlighten*. 2nd ed. El Dorado Hills, CA: Analytics Press, 2012.

Books about the ethics of data:

- O'Neil, Cathy. *Weapons of Math Destruction: How Big Data Increases Inequality and Threatens Democracy.* New York: Broadway Books, 2016.
- Broussard, Meredith. *Artificial Unintelligence: How Computers Misunderstand the World.* Cambridge, MA: MIT Press, 2018.
- Eubanks, Virginia. *Automating Inequality: How High-Tech Tools Profile, Police, and Punish the Poor.* New York: St. Martin's Press, 2017.

Finally, if you want to learn more about the charts that appear in this book, visit http://www.howchartslie.com.

Index